T0231224

Tractus Immuno-Logicus
A Brief History of the Immune System

Antonio A. de Freitas, MD, PhD
Professor at the Institut Pasteur
Paris, France

CRC Press is an imprint of the
Taylor & Francis Group, an **informa** business

To my parents, Antonio & Corina

⋅⋅⋅ Contents ⋅⋅⋅

✦✦✦ **Preface** ✦✦✦

The history of this text started years ago after reading Wittgenstein's "*Tratactus Logico-Philosophicus*". At some time later, it seemed to me a good idea to follow the "*tratactus*" structure to attempt to write a minimal description of the immune system. I finally did it for fun and hopefully to be useful to whomever reads it.

The text reflects my own personal view of the vertebrate's immune system (IS). It is centered on concepts and ideas that were developed since 1986 based on work from my own lab[1] and from Benedita Rocha's lab[2] and I'm greatly indebted to her for this. I have kept it short and focused on what I believe are the essential features of the IS. I've tried to avoid too much detail and most of the complex immunology jargon. If some now fashionable aspects of the IS are only superficially mentioned it is because I feel that they may be not so relevant after all. Perhaps for all these reasons I give no detailed sources and simply refer the reader to some general inspiring non-immunological references.

The reader may find the outline structure unusual for a scientific text. The text is divided in different sections. In each section, statements are numbered by order of relatedness and importance creating a kind of hierarchical affiliation. For example, statements 2.1.01 and 2.1.1 support or comment on statement 2.1, but statement 2.1.01 is "closer" or more related to 2.1 than statement 2.1.1. Each series of 2.1 statements should be read independently i.e., as a paragraph. However, relationships between the numbers are not strictly hierarchical and concepts are sometimes repeated. Numeration is used to help define concepts that are important to understand as stated. Spacing is used to create breaks and facilitate reading. The text can be read either in one go or spaced according to numeration and the layout of the text. Each section can be read on its own.

I look forward to raising in the general non-scientific reader an interest for an immune system where lymphocytes are mainly "concerned" with replication, survival factors and homing to the appropriate niche. This is the 1960s "sex, drugs and rock'n roll" view of the IS. Moreover, there are many concepts that are shared with other fields, e.g., ecology, economics.

I hope to stimulate quite a lot of discussion among those that study the Immune System. The text opens opportunities on Immunology teaching by focusing on concepts, interactions and their relatedness and all those as one. The readers may build frameworks of cross-references between statements that are not in line to create alternative reading paths. They should interact with each other to compare interpretations and refer to the immunology literature. They may create new connections, add new sub-sections, references and suggest modifications. The reader is encouraged to send their comments, corrections, additions, etc. to antonio.freitas@pasteur.fr.

To the medical doctor or the advanced specialist the text encapsulates the Immune System and provides a novel prism with which to approach Immunology. By attempting to always follow a logical line of thought, I end up by making new statements, some of which remain hypothetical, waiting for experimental testing, that change the current views of the IS. The purpose was that "each" statement should force the reader to stop, think and whenever possible, test. By doing so, I hope to provoke new questions and inspire new experimental approaches and research.

While working on this manuscript and looking for inspiration, I played many games of Shanghai II. Sometimes I got the cookie "Wise men learn much from fools..." There are many "fools" in science. Dear reader, please be wise.

Antonio A. de Freitas, MD, PhD

[1] Freitas AA, Rocha B, Coutinho AA. *Immunol Rev* 1986; 91:5. Freitas AA, Rocha B. *Immunol. Today* 1993; 14:25. A. A. Freitas AA, Rosado MM, Viale AC, Grandien A. *Eur J Immunol* 1995; 25:1729. Rocha B, Grandien A, Freitas AA. *J Exp Med* 1995; 181:993. Freitas AA, Agenes F, Coutinho GC. *Eur J Immunol* 1996; 26:2640. McLean AR, Rosado MM, Agenes F, Vasconcellos R, Freitas AA. *Proc Natl Acad Sci USA* 1997; 94:5792. Tanchot C, Lemonnier FA, Perarnau B, Freitas AA, Rocha B. *Science* 1997; 276:2057. Agenes F, Rosado MM, Freitas AA. *Eur J Immunol* 1997; 27:1801. Agenes F, Freitas AA. *J Exp Med* 1999; 189:319. Freitas AA, Rocha B. *Annu Rev Immunol* 2000; 18:83. Almeida AR, Borghans JA, Freitas AA. *J Exp Med* 2001; 194:591. Gaudin E, Rosado M, Agenes F, McLean A, Freitas AA. *Immunol Rev* 2004; 197:102. Almeida AR, Rocha B, Freitas AA, Tanchot C. *Semin Immunol* 2005; 17:239. Hao Y, Legrand N, Freitas AA. *J Exp Med* 2006; 203:1643. Almeida AR, Zaragoza B, Freitas AA. *J Immunol* 2006; 177:192.

[2] Rocha B, Dautigny N, Pereira P. *Eur J Immunol* 1989; 19:905. Rocha B, von Boehmer H. *Science* 1991; 251:1225. Tanchot C, Rocha B. *J Exp Med* 1997; 186:1099. Tanchot C, Rocha B. *Immunol Today* 1998; 19:575. Veiga-Fernandes H, Walter U, Bourgeois C, McLean A, Rocha B. *Nat Immunol* 2000; 1:47. Peixoto A et al. *J Exp Med* 2007; 204:1193. Rocha B, *Immunol Rev* 2007; 215:166.

⋅⋅⋅ Acknowledgements ⋅⋅⋅

I thank Benedita for sharing and inspiration.

My parents for letting me be.

Filipa for what was and what is yet to come.

I thank Drs. Wendy Haston and Robert Wildin for encouragement, discussion and valuable suggestions; Drs. Adrian Hayday and Alfred Singer for helpful comments; Drs. Maria De Sousa, Fréderic Jacquemart and Geneviève Milon for suggestions; Drs. Marc Daeron, Sylvie Garcia and Peter Oppenheimer for reviewing the text.

I thank the members of the "Laboratoire des Dynamiques Lymphocytaires" and of the "Unité Biologie des Populations Lymphocytaires" and in particular Angela McLean, Afonso Almeida, Manuela Rosado, Nicolas Legrand and Fabien Agenes for their help.

In the course of the years I met quite a few students and post-doctors. Some I did not have to teach and I just let them remember what they already knew.[3] They taught me a lot instead.

I thank Ronald Landes for daring to publish an "unusual" book and Cynthia Conomos and Celeste Carlton for help with the production.

Finally, I must acknowledge the input of many different contributors whether willingly or not intentionally.

[3] After Kierkegaard "Philosophical Fragments".

I. The Earth is round.

II. This is "all" that is important to know about Earth's geography.

III. All the rest are details.[4]

---

[4] Adapted after N.K. Jerne commenting about what he called the ultimate experiment: the circumnavigation travel by Fernão de Magalhães proving that the earth was round.

Chapter 1. Definition

1 The immune system is a set of rules (interactions) rather than a cohort of cells.

1.1 That is, it is a set of interactions that defines the immune system.

1.1.2 Since these interactions determine response or non-response,

1.1.2.1 the decision of an immune response implies a non-response may also be decided.

◆◆◆

1.2 Rules determine dynamics (succession of states),

1.2.1 the cohort of all cells and all their rules define dynamics.

1.2.2 It is essential for a cell to be part of a dynamic—a "state".

1.2.2.1 A state that includes all aspects of past history relevant for future probabilities.

1.2.3 A cell can only be part of a state if such possible state is coherent to the cell.

◆◆◆

Chapter 2. Evolution

2.1 Self-replicating molecules define life.

2.2 Self-replicating molecules can only develop, persist and replicate in favorable environments.

2.2.1 Self-replicating molecules must ensure their own survival through the control of their immediate environment.

◆◆◆

2.3 A self-replicating unit is an ensemble of self-replicating molecules that developed a surrounding protected microenvironment.

2.3.1 Within a self-replicating unit, DNA self-replicating molecules may exist in two modes: active and silent.

2.3.1.1 When active they code for the production of molecular structures and pathways that contribute to shape their immediate environment in order to survive.

2.3.1.2 In the process, they are first transcribed into intermediate RNA molecules that are thereafter translated into protein molecules.

2.3.1.2.1 Both transcription, translation and protein products must be tightly regulated to optimize the survival of the self-replicating unit.

2.3.1.3 Each unit survives and replicates as a whole.

Tractus Immuno-Logicus: A Brief History of the Immune System
by Antonio A. de Freitas
©2009 Landes Bioscience

2.3.2 Self-replicating DNA molecules mutate since replication is error-prone.

2.3.3 Mutation may either hinder or allow self-replicating molecules to survive.

2.3.3.1.1 Since mutations may be harmful, at the primary molecular level, the fate of an isolated single molecule is uncertain and determined by chance.

2.3.3.1.2 Thus, each molecule (has) a probability of replicating successfully.

2.3.4 The association of different self-replicating molecules reduces the uncertainty of survival for each molecule.

2.3.4.1 It promotes the arousal of structures that protect the immediate microenvironment from the exogenous environment.

2.3.4.1.1 It also permits the development of mechanisms that repair damaged replicating molecules.

2.3.4.2 Association of different isolated molecules within a unit overrides the individual probability function of each separate molecule and increases their mutual probability of survival.

2.3.5 Necessity imposes that the capacity to generate a protected microenvironment is "preimprinted" in the surviving self-replicating molecules.

2.3.5.1 After selection, uncertainty is no longer apparent among the existing self-replicating units.

2.3.5.2 Each unit is selected as a whole.

2.3.6 Thus, uncertainty applies at smaller dimensions e.g., to isolated single molecules and is mitigated in larger units assembling several selected self-replicating molecules.

2.3.6.1 That is to say: selection of those molecules, whose ensemble enhances survival, makes the underlying uncertainty less apparent.

◆◆◆

2.4 Single cell protists are self-replicating units.

2.4.1 A single cell protist only survives and replicates in a favorable environment.

2.4.2 A favorable environment supplies sufficient quantities of factors required for survival and replication ("resources").

2.4.2.1 According to their environment single cell protists may die or survive.

◆◆◆

2.5 The composition of the environment determines the selection of the cells capable of surviving.

2.5.01 Varying environments result in the survival of different cells with different survival requirements.

2.5.1 Through the consumption of resources, single cell protists change their surrounding environment.

2.5.1.1 By modifying their surroundings, single cell protists participate in their own survival

2.5.1.2 and in the selection of other cells with different requirements.

◆◆◆

2.6 The variable composition of the environment favors emerging originality and diversity.

2.6.1 Random mutations provoke the emergence of single cell associations that favor mutual survival. This process allowed the development of multicellular organisms.

2.6.1.1 In a cell association, during periods of shortage, dying cells may provide ready-made nutrients to neighboring cells capable of reusing them.

2.6.1.2 In these types of association, cells lines establish symbiotic interactions each with distinct functions to assure the transmission of the replicating molecules from both partners.

2.6.1.3 Within a multicellular organization, cells may seem to function autonomously or collaboratively at different levels.

◆◆◆

2.7 Transition from single cell to complex multicellular individuals implies that the cells are also subjected to selection within the somatic internal environment.

2.7.1 This implies that a hierarchical organization is established, the proliferation and survival of the cells from each different line is strictly controlled and their number kept constant.[5]

◆◆◆

---

[5] "A hierarchical organization is established, the replication and survival of each different cell line restrained and strictly controlled and the number of individuals in each cell line kept constant" L.W. Buss.

2.8 Various control systems adjust interactions with the surroundings to ensure that the composition of the protected internal microenvironment remains constant.

◆◆◆

2.9 The cohort of all cells and all their interactions evolve to reach steady state dynamic equilibrium, that is, homeostasis.[6]

2.9.1 Homeostasis permits complex organisms to adapt to multiple changes in their surroundings and survive.

2.9.2 Homeostasis is an intrinsic property resulting from cell assembly interactions with its immediate environment and is characteristic for each different cell assembly.

◆◆◆

2.10 Homeostasis of a population of cells can be defined as "a return tendency, due to density-dependent processes, to approach a stationary distribution of population densities", numbers.[7]

◆◆◆

2.11 Homeostasis is also the dynamic equilibrium of all components of a living organism.

◆◆◆

[6] Claude Bernard and W.B. Cannon.
[7] After I. Hanski.

2.12 Self-replicating molecules, single cell protists and multicellular individuals follow the same basic rules for survival and replication.

2.12.1 The presence of a similar set of rules acting at different levels of complexity preserves symmetry in biological systems.

2.12.1.1 Within cells, interactions between molecules follow the same sets of rules.

2.12.2 Thus, complex organisms and simple cell protists use basically identical strategies to survive and replicate.

2.12.3 Complex organisms develop body praxis that although using basically identical strategies require hierarchies introducing asymmetries that serve the overall preservation of symmetry.

2.12.4 Asymmetry is essential for life by introducing chemical imbalance and diversity of environments.

◆◆◆

2.13 A cell can only be part of a multicellular organism if such possible dynamic ("state") is "pre-imprinted" in the cell.

2.13.1 Cells from multicellular organisms are independent in that they can occur in all possible situations, but only if they are part of a dynamic state, the organism, thus a form of dependence.

2.13.1.1 In these conditions a mature cell isolated from a multicellular cohort is "nothing" by itself, its only value is the dynamics it helps establish.

2.13.2 In multicellular organisms, the "coherent" organization and the state of dependence—"homeostasis"—of the individual cell lines must be preserved. Otherwise the "state" and its parts cease.

2.13.2.1 If a cell is unable to find its "repairs" and is unable to be coherently part of a dynamic it may enter a program of cell death—apoptosis.

◆◆◆

2.14 To know a cell is to know all its possible "states".[8]

2.14.1 To know a cell is to know all its properties.

◆◆◆

2.15 Different cell states result from the activation of different sets of replicating molecules.

2.15.1 Cells may change states according to their dynamics and environmental cues,

2.15.1.1 meaning that different sets of replicating molecules can be activated and change states in response to environmental signals.

◆◆◆

2.16 The development of multicellular organisms from their single precursor cell follows a preestablished program that occurs through multiple steps of differentiation.

---

[8] Which is practically impossible.

2.16.1 Each step of differentiation implies the activation of new replicating molecules (genes) and the silencing of others, to generate new cell lines with different functions—new roles.

2.16.1.01 The activation or silencing of selective genes is a probabilistic event and each individual cell will follow a unique decision path, which will not be perceptible at a cell population level.

2.16.1.02 Only those cells that follow the "accurate" path survive.

2.16.1.03 The accurate path being defined by the ability of the resulting cells to be part of a dynamic "state" e.g., organism.

2.16.1.04 Cells growing autonomously out of control destroying the host.

2.16.2 Fine-tuning of gene expression is achieved by the balance between inhibitory and enhancing regulatory elements.

2.16.2.01 Due to their importance and the complexity of the interactions necessary to maintain states in complex organisms, these regulatory elements may represent the majority of replicating molecules in the genome.

2.16.2.02 Small interfering RNAs that inhibit or enhance gene product expression play a major role in regulation.

◆◆◆

2.17 Development reflects interactions between different cell lines in the quest for survival and replication.

2.17.1 Each new cell line generated may act to limit replication of another, while enhancing its own.

2.17.2 Cell lines will also provide factors and structures necessary for the survival of other cell lines.

2.17.3 These mutualistic types of interactions are determinant for the organization of complex multicellular organisms.

◆◆◆

2.18 During development cells transit from a multipurpose state to a state of specialization and individuality.

2.18.1 By specialization, cells avoid conflict with other cells

2.18.2 and by acquiring new individual properties they also contribute to the survival of the whole.

◆◆◆

2.19 In the transition from a multipurpose progenitor to a specialized function, each cell line creates or finds its own surviving and replicating niche.

2.19.1 Only by adapting to new niches can the different cell lines ensure their individuality and survival against the conflicting interests of other cells.

◆◆◆

2.20 Development proceeds to its final stage when all different cell lines reach equilibrium in their niches and in the whole.

2.20.1 Cell proliferation is limited to the Hayflick's number,[9] i.e., a single human cell cannot undergo more than 50 cell divisions to give rise to more than 10^{15} cells (approximately 1 ton of tissue mass).

2.20.1.01 The Hayflick's limit varies according to the longevity of the species (it is higher for cells from long-lived turtles) and is reset to zero during meiosis.

2.20.1.1 The Hayflick's limit is reached when "non-renewable resources" required for cell division become limiting as cell division progresses.

2.20.1.2 The alterations associated with progressive cell division together with changes in entropy values lead to the role of time in cell development.

2.20.2 Thus, the modifications that lead to cell specialization through proliferation and differentiation follow the arrow of time and are therefore irreversible.

2.20.2.01 The rules that apply to a single cell protist will also apply to the cells of a multicellular organism, but the reverse is not true: the rules that are followed by cells of a multicellular organism will not necessarily apply to single cell protists.

◆◆◆

2.21 Cooperation and conflict interactions between individual cell lines collectively contribute to the survival of the whole and thus, ultimately, to the propagation of self-replicating molecules.

◆◆◆

---

[9] Leonard Hayflick.

2.22 As all individual cell lines of a multicellular organism derive from a single progenitor cell it can be said that they are entangled.

2.22.1 They share interactions and a coherent organization.

◆◆◆

2.23 Complex organisms host multiple cell lines fulfilling different needs that contribute to the survival of the whole and to their own survival.

◆◆◆

2.24 They provide environments with ready-to-use resources favorable to the survival of different types of cell lines.

2.24.01 They secure somewhat a relative freedom from the external environment.

2.24.1 Foreign cell lines or small replicating molecules may invade complex organisms to use these resources to their own benefit.

2.24.2 This interaction may bring mutual benefits to both host and invader, and lead to their mutual adaptation,

2.24.2.1 or compromise the survival of either.

2.24.3 In the later case either the host eliminates the invader and survives,

2.24.3.1 or the invader finds a new host to ensure the survival of its replicating molecules.

◆◆◆

2.25 In a vertebrate, different cell lines specialize in interacting with alien invaders,

2.25.01 to ensure either protection against noxious invaders and neutralize their harmful effects or facilitate their safe integration in the host—symbiosis,

2.25.02 and enhance the survival of the host and of its replicating molecules.

2.25.1 As the rate of replication of simple molecules or single cell protist is faster than that of complex cell lines, hosts evolved ready-to-use mechanisms that rapidly sense and interact with the invader to protect the host.

2.25.1.001 These include interactions of invader associated molecular patterns with pattern recognition receptors and variable host's cell surface structures with leucine rich repeat segments.

2.25.1.01 Besides there are also basic cellular and soluble mechanisms that prevent host invasion and dissemination of invaders—integrity of epithelial barriers, secretion of factors that block spreading and proliferation of invaders, etc.

2.25.1.02 Commensal bacteria contained in well-defined niches behave as non-invader symbionts. However, outside their niche they may sometimes act as harmful invaders and be treated as such.

2.25.1.1 The immediate interactions that can neutralize invaders are inherited through evolution and reminiscent of primordial mechanisms on which relies the perpetuation of simple protist organisms in their flight for survival. It is said that they are part of the "innate" defenses.

2.25.2 Because invaders evolve and change, vertebrate hosts have also developed efficiency-increased mechanisms to interact with their potential invaders and adapting to new challenges.

2.25.2.1 Innate mechanisms that protect short-lived invertebrates, however, do not suffice to protect long-lived vertebrates.

◆◆◆

2.26 The set of "rules" (cell interactions) that allow vertebrates to deal with noxious foreign invaders constitutes the immune system (IS).

2.26.01 Invaders that induce biological processes has that compromise survival of the host are pathogens.

2.26.02 Vertebrates without an IS fail to survive in unprotected environments.

2.26.1 Each organism evolved the type of IS that fits its needs for survival.

◆◆◆

2.27 As invaders come in many different shapes the IS must be able to interact with (recognize) all possible invaders.

◆◆◆

2.28 A cohort of cell lines of the IS (immune-competent cells) specializes in the interaction with the alien invaders (recognition of) or with the changes that they induce in the cells and tissues of the host.

2.28.1 These immune-competent cells display at their surface, molecules capable of interacting with a multitude of different molecular shapes.

2.28.2 These cell surface molecules are termed "receptors", and their interactions with other structures are termed "recognition".

2.28.2.1 It should be pointed out, however, that these receptors cannot discriminate what is invader from what is host. They simply interact with particular structures.

◆◆◆

2.29 The ability of the IS to recognize infinite molecular forms is encoded in the genome of the host,

2.29.01 thus determining its "Promethean" view (capacity to predict and recognize new non-existing structures).

◆◆◆

2.30 During development, the cells of the IS generate a diverse array of receptors by the random recombination and association of different gene segments.

2.30.1 These operations are conducted by Recombination Activating Genes (RAG) that code for enzymes specialized in gene segment recombination.

2.30.2 RAGs emerged late in evolution and were determinant for the core structure of the adaptive IS in vertebrates.

2.30.2.01 RAGs are the end-point of the strategies developed through evolution to achieve the anticipatory "Promethean" recognition of all universal molecular forms.

◆◆◆

2.31 RAGs induce the random recombination of different gene fragments that code for the recognition (V-variable) site of the antigen-specific receptors present on the surface of lymphocytes.

2.31.01 A lack of precision in the association made between the different genetic elements during recombination, generates further diversity.

2.31.1 These mechanisms of generation of diversity bestow the IS with a repertoire of receptors able to recognize a universe of different shapes.

♦♦♦

2.32 Recognition is degenerate, i.e., each receptor can recognize many different shapes and the same shape can be recognized by several distinct receptors.

2.32.1 This property further increases the capacity of the IS to cover all possible existing or non-existing shapes.

♦♦♦

2.33 As invaders act in different ways, a wide variety of host strategies deal with each type of invasion.

2.33.1 Some of the host cell lines act directly against the invader, while others act against host cells harboring the invader or its replicating molecules.

2.33.2 Some cells of the IS—lymphocytes—express at their surface structures capable of interacting with specific 3-dimensional (3D) molecular shapes—antigen specific receptors.

2.33.2.1 These receptors exert a dual function: trough their variable regions they allow recognition of 3D molecular shapes and once engaged their constant regions signal their bearing cell to initiate a kinetic process that we call response.

2.33.2.2 These receptors can interact with all possible shapes present in the environment, i.e., either intrinsic to the host or belonging to the invader.

2.33.2.3 The number, local density of bound receptors and the type of interacting environment determine either the non-response or the type of response of the cell.

◆◆◆

2.34 When the cell lines act directly against the invader (antigen), this implies direct recognition of the antigen by the immune-competent cell.

2.34.1 The cell lines (B lymphocytes) that directly recognize the invader ("antigen") specialize in the production of soluble molecules ("antibodies") that exert also a dual function: recognition of the invader and elimination of the invader.

2.34.1.1 The antibodies produced have the same recognition site as the recognition receptor present at the surface of the B cell.

2.34.2 This functional dichotomy of the antibody (immunoglobulin) molecule is discernible in the genetic organization of its genes.

2.34.2.1 Some gene fragments code for the recognition (V-variable) site while others code for the constant (C) effector site of the immunoglobulin (Ig).

2.34.2.1.01 An immunoglobulin molecule comprises a heavy chain (H) and a light chain (L) polypeptide, each with a variable and a constant region.

2.34.2.1.02 The association between the V regions of the two chains forms the antigen-binding site.

2.34.2.1.02.1 The random association of the two chains further contributes to the generation of diversity.

2.34.2.1.03 B cells of the same progeny—clone—maintain the same H- and L-chain V-regions that form the antigen interaction sites, but may express different Ig constant parts.

2.34.2.1.03.1 The presence of dissimilar constant parts defines different Ig (iso)types and corresponds to distinct "effector" functions.

◆◆◆

2.35 Cell lines (T lymphocytes) that act against host cells harboring invaders specialize in the production of toxic molecules that kill the infected target cells and thus, help to prevent the invader from spreading.

2.35.1 T cells express receptor molecules (TCR) that like immunoglobulin (Ig) are formed by two chains ($\alpha + \beta$ or $\gamma + \delta$) each with a variable and a constant region.

2.35.1.1 The antigen-binding site is again formed by the association between the V regions of the two chains.

2.35.2 Alien invaders may remain as autonomous entities in the host tissues or invade the host cells loosing their autonomy.

2.35.2.1 In the first case, products of the invader may be present in the host cells.

2.35.2.2 In the second case, products of the host cell are modified or associated with parts of the invader.

2.35.3 These processes oblige the fitted immune-competent T cell to interact with (recognize) both structures of the host cells and of the invader.

2.35.4 A cell cannot, intrinsically, discriminate what is invader from what is host. It simply interacts with a particular structure.

2.35.4.1 As the intruder changes the host and the cells it infects, the immune cell interacts with (recognizes) the modified structures of the compromised cell.

2.35.4.1.1 Cells of the host express sets of molecules that are designed to be selectively associated with products of the invader. They constitute the major histocompatibility complex (MHC).

2.35.4.1.2 It is said that these structures "present" products of the invader.

2.35.5 The diverse antigen-specific receptors expressed by the T cells (TCR) preferentially interact with these specialized MHC structures present in the hosts' cells.

2.35.5.01 T cells interact fragments of the invader (peptides) associated with MHC molecules—they "recognize" peptide/MHC complexes.

2.35.5.1 Polymorphism of the MHC structures ensures that at least some individuals within the species will have the appropriate set of molecules to interact and be modified by new lethal invaders.

2.35.5.1.1 In absence of the "correct set" of modifiable molecules, T cells may not be able to interact with the invader and protect the host.

2.35.5.1.1.01 If this were the case the host will die.

2.35.5.1.2 Polymorphism serves to guarantee the preservation of the species.

2.35.5.2 Thus, as each host has its own distinct set of molecules to exert these functions, the MHC also acts as an "identity" marker.

2.35.5.2.1 This implies that in each host the TCR and the "identity" markers must match.

2.35.6 T cells interact with changes of these host "identity" markers.

2.35.6.1 These changes either imply intrinsic changes of host molecules or the presence of extrinsic parts of the invader itself presented on the surface of the infected host cell.

2.35.6.2 Hosts evolved two different types of pathways and markers to deal with "intrinsic" or "extrinsic" modifications.

2.35.6.2.01 The host has evolved a special set of natural killer (NK) cells to eliminate the infected cell where the intruder suppresses expression of the MHC molecules.

◆◆◆

2.36 The ability of the identity markers to specifically signal the presence of the invader and of the cells of the IS to selectively interact with the identity marker and its alterations must be simultaneously present in each host.

◆◆◆

2.37 This state of cell-to-cell interactions (recognition) can only occur, if such possible states pre-exist in the host.

2.37.1 Or, as is the case, they co-evolved in all the hosts of the same species.

◆◆◆

2.38 Similar cell-to-cell interactions operate in all protist organisms.

2.38.1 And they are required for the harmonious development of multicellular organisms.

2.38.1.1 Thus, in the absence of such "recognition" events perpetuation of a species is not possible.

◆◆◆

Chapter 3. Development

3 Lymphocytes develop from hematopoietic multipurpose progenitor cells in defined host sites ("primary" lymphoid organs).

3.1 During the host's ontogeny and life span, both the site of lymphocyte origin and the characteristics of the progenitor cells change.

3.1.1.01 Lymphocyte development moves from the embryonic yolk sac and fetal liver to the adult bone marrow and thymus.

3.1.1.02 With time there are also quantitative changes as lymphocyte production first increases early in ontogeny and then decreases with aging.

◆◆◆

3.2 Lymphocyte development follows a dynamic program that occurs through multiple steps of differentiation and proliferation.

3.2.1 It involves a sequential set of events, each occurring with a measurable probability and whose description is necessary,

3.2.1.01 but where the isolated identification of separate multiple progenitor populations may turnout to be irrelevant.

3.2.2 Each step implies the activation or silencing of selected genes randomly,

3.2.2.1 resulting, according to the patterns of genes expressed, in distinct cell populations with different functions.

3.2.3 A fraction of the progenitor population must retain its multipurpose capacities to sustain further lymphocyte generations necessary to cover the host's life span.

3.2.3.1 This implies a mechanism whereby after cell division one daughter cell retains the properties of the progenitor while the second diverges and acquires new properties.

3.2.3.2 This process may be achieved by either asymmetric cell division or by cell seeding.

3.2.3.2.1 The rate of asymmetric versus symmetric division and differentiation is determined by the host's dynamics.

3.2.3.2.1.1 Following depletion of matured cells, the rate of proliferation among the progenitor cells is higher and the rate of asymmetric division proportionally lower.

3.2.3.2.2 Overgrowth of the population of dividing progenitors may push them out from the original environment to a new environment, where they are driven to differentiate.

3.2.3.3.3 When two daughter cells end up in different microenvironments, this may mimic asymmetrical cell division and lead to different cell lineage fates.

3.2.4 During development, activation of particular genetic elements selectively commits precursor cells to differentiate into distinct cell lines.

3.2.4.01 The initiation of new gene expression is stochastic. Each cell may start to express the same gene at different times and at different levels.

3.2.4.02 Gene expression is triggered by enhancer factors that act by increasing the probability of transcription.

3.2.4.03 Thus, multiple transcription factors and other regulatory proteins, each with their own activation pathway controlled by previous stochastic events, regulate gene expression.

3.2.4.04 Such initial variation of gene expression generates further diversification. Thus, a homogeneous progenitor population may generate very different cell types through successive diversification steps.

3.2.4.05 These cells may be more constrained in their differentiation capacity.

3.2.4.1 Commitment of precursor cells to a particular differentiation pathway—fate—is induced by environmental signals.

3.2.4.1.1 The primary lymphoid organs are unique as they contain the environmental factors required for both maintenance of multipurpose progenitors and lymphocyte commitment and differentiation.

3.2.4.1.1.1 Indeed, multicellular organisms generate anatomical structures that permit other cell lineages to develop.

3.2.4.2 Upon commitment, cells follow one of several pre-imprinted programs that will allow their survival in new environments.

3.2.4.2.01 Stochastic expression of the program confers plasticity to the differentiation process and contributes to diversity of the resulting cell populations increasing their chances of survival.

3.2.4.2.1 Simultaneously, at each step of differentiation, cells adjust their anatomical localization within the primary organ by the expression of different receptors for local attractants.

3.2.4.2.2 During this migration, changes in the surrounding cellular microenvironment further influence lymphocyte differentiation.

3.2.4.3 The multipurpose capacity of the progenitor cell becomes progressively restricted, thereby reaching a state of specialization and individuality.

3.2.4.4 On the whole, this is an irreversible linear process determining an arrow of time, with cells acquiring at each step of differentiation new characteristics that will allow them to exert a specialized function.

◆◆◆

3.3 Characteristic of lymphocyte development is the acquisition of cell surface structures capable of interacting with a universe of shapes.

3.3.01 That is to say: they express receptors capable of recognizing multiple antigens.

3.3.1 RAG proteins are determinant for the synthesis and expression of the antigen specific receptors in B and T lymphocytes.

3.3.1.1 RAG proteins are selectively expressed by lymphocyte precursor cells at some specific points of their development.

3.3.1.2 By cleaving DNA at specific sites they, together with DNA repair enzymes, rearrange gene segment organization and allow the generation of multiple diverse antigen receptors.

3.3.1.2.1 Gene rearrangement occurs stepwise in a probabilistic manner and cells that fail one step may not survive as they fail to interact with the environment— they cannot be part of a "state".

3.3.1.2.2 Because lymphocyte development requires the intervention of multiple factors acting at specific stages of differentiation in a concerted manner,

3.3.1.2.3 lack of some of these factors or changes in their sequential activation leads to selective cellular defects of the immune system (IS).

3.3.1.2.4 The IS may lack particular cell subsets or fail to exert specific functions.

3.3.1.3 In extreme cases, e.g., defects in RAG expression or activity, hosts lack all lymphocytes.

3.3.1.3.1 Hosts without lymphocytes survive, if kept in protected and isolated environments.

3.3.1.3.2 This shows that lymphocytes are not strictly required for the normal development of the host,

3.3.1.3.2.01 which does not mean that when they are present they do not modify the body's economy.

3.3.1.3.2.02 They can therefore be used to alter and control other body functions, e.g., endocrine.

3.3.1.3.3 It also shows that the IS is critical for the interactions of the host with the surrounding environment.

3.3.2 In their ability to recognize a universe of antigens, lymphocytes differ from the other cell lines of the host,

3.3.2.1 they occupy a unique niche and exploit unique resources—antigens.

3.3.2.2 Antigens act as resources because they favor lymphocyte survival and/or division within some range of their availability.

3.3.2.2.01 This means that resource availability outside critical values may either be inactive or, when in excess, deleterious to lymphocyte survival.

3.3.3 At particular early differentiation stages, B and T cells also display incomplete antigen recognition receptors.

3.3.3.01 These pre-receptors have a limited recognition capacity and determine if the cell is "coherent" to continue differentiation or not—if it can be part of a "state".

3.3.3.02 If the receptor is coherent (if it can capture survival signals) the cell continues receptor rearrangement and differentiation.

3.3.3.03 If the receptor is not coherent there is still a probability for the cell to change and attempt to ensure survival by rearranging and expressing a new receptor.

3.3.3.1 Lymphocyte development is therefore a highly wasteful process with considerable cell loss at each step.

3.3.3.2 Only the fitted cells surviving.

3.3.4 Different populations of lymphocytes express unique markers that permit their identification.

3.3.5 B cells are characterized by the surface expression of Ig that act as antigen-specific B cell receptors (BCR) and exerts a unique shape recognition function.

3.3.5.01 In adult birds, B cells develop in a specialized organ—the Bursa of Fabricius—thus the name B cell.

3.3.5.1 According to the age of the host and the site of origin, B cell subsets with different characteristics evolve, environments dictating the type of development.

3.3.5.1.1 One may distinguish two major subsets of B cells—B1 and B2—that have different progenitor origin, location and dynamics.

3.3.5.2 Besides BCR, B cells express CD19, a signal transducer receptor selectively expressed by the cells of the B lineage, and the Igα/Igβ invariant chains associated with and specialized in translating the signals captured by the BCR.

3.3.5.3 Development of B cells implies a process of selection where only those cells that by chance recombine and express a functional BCR can complete differentiation and survive.

3.3.5.3.1 For the B cell to survive its BCR should ensure capture of specific antigen resources inside critical ranges. Too little may not suffice and too much may be deleterious to lymphocyte survival.

♦♦♦

3.3.6 T cell development occurs in a specialized primary organ—the thymus—thus the name T cell.

3.3.6.01 The thymus environment is able to induce full commitment of progenitor cells to T cell development.

3.3.6.02 The thymus environment is progressively modified by interactions between the thymus epithelium, the thymus stroma and the developing lymphocytes. It provides distinct anatomical sites required for specific steps of T cell development.

3.3.6.03 T cell lineage fates seem to be determined by signals that involve or control the Notch signaling pathway; Notch signals induce T cell commitment.

3.3.6.1 T cells are characterized by the expression of antigen-specific cell receptors (TCR) that preferentially interact with the "identity" markers of the host cells and its modifications.

3.3.6.1.01 In absence of invaders the vertebrate identity markers are associated with own products of the host's cells.

3.3.6.1.1 The identity markers of the host are called antigens of the "Major Histocompatibility Complex" (MHC) and vary from host to host.

3.3.6.1.2 The ability of the T cells to selectively interact with the host's identity markers is "pre-imprinted" in all the hosts of the same species.

3.3.6.1.3 Thus, in each host, development of T cells implies a process of selection whereby only the T cells that by probability recombine and express a TCR capable of interacting with the identity markers of the host can survive.

3.3.6.1.4 As the ability of the T cells to interact with the different host's identity markers is pre-imprinted in all the hosts of the same species, there is a high probability that among the surviving mature T cells some will also interact with and react to identity markers of other individuals of the same species as if they were modified host molecules.

3.3.6.1.4.01 By developing out of context, these cells, capable of interacting with identity markers present in other individuals, will take their presence as signals of invasion and react accordingly.

3.3.6.1.4.02 The presence of these "alloreactive" T cells prevents the exchange of cells and tissues between individuals of the same species since they can recognize and eliminate tissues from disparate individuals (transplants).

3.3.6.2 There are two types of TCRs expressed by different T cell populations: the αβ and the γδ TCR.

3.3.6.2.1 Development of αβ and γδ T cells diverges quite early in the thymus.

3.3.6.2.2 αβ and γδ T cells differ in recognition patterns, function and location.

3.3.6.2.2.1 The αβ T cells are primarily found in lymphoid structures while the γδ T cells are located mainly at the borders of the body economy with the exogenous environment.

3.3.6.2.2.1.1 The γδ T cells recognizing specific tissue markers to allow invader localization.

3.3.6.3 T cell development also occurs in the gut epithelium, which shares the same endoderm origin with the thymus epithelium.

3.3.6.3.1 These intra-epithelial αβ T lymphocytes (IEL) exert unique functions being located at the border with the immediate exogenous environment.

3.3.6.4 T cells also express CD3 a complex of several invariant units associated with the TCR chains and that translates signals captured by the TCR,

3.3.6.4.1 and CD4 or CD8 co-receptor molecules that participate in cell-to-cell recognition interactions.

3.3.6.5 T lymphocytes use the CD4 and CD8 co-receptors to facilitate recognition of the two main classes of host's identity markers.

3.3.6.5.01 CD4 and CD8 co-receptor expression determines the patterns of αβ TCR recognition and its restriction to the MHC.

3.3.6.5.1 Hosts express two types of identity markers, dealing with different types of invasion,

3.3.6.5.1.1 and different T cells recognize each type of identity marker.

3.3.6.6 Products of the invader may be captured by the host cells and re associated with MHC class II molecules,

3.3.6.6.1 or the host cell is modified by the infection to make parts of the invader associated with or presented by the MHC class I molecules.

3.3.6.7 CD4 help T cells to interact with MHC class II identity markers that present parts of the invader.

3.3.6.7.1 CD8 help T cells to interact with MHC class I identity markers that are associated with parts of the invader.

3.3.6.7.2 Early in differentiation, T cells express both CD4 and CD8.

3.3.6.8 During development the differentiating lymphocytes integrate different trophic signals and only those T cells expressing a coherent pattern of recognition by the TCR and the CD4 or CD8 co-receptors complete maturation and survive.

3.3.6.8.1 Developing cells that either fail to connect or that establish too strong interactions abort their development.

3.3.6.8.1.1 In the thymus the cellular environment ensures expression of particular sets of genes, e.g., AIRE, Auto-Immune Regulation, which determine transcription of organ specific genes and contribute to modulate T cell development.

3.3.6.8.2 CD4 and CD8 T cell lineage[10] choice is determined by the strength and persistence of TCR interactions with MHC ligands in the thymus, control of co-receptor and cytokine receptors and expression of specific transcription factors like Runx3, which induces CD8 T cell lineage choice.

3.3.6.8.3 At the end, two major classes of αβ T lymphocytes are formed: CD4 T cells, which recognize MHC class II identity markers, and CD8 T cells, which recognize MHC class I identity markers.

---

[10] Cell lineage: a genealogic pedigree of cells related through mitotic division.

3.3.6.8.3.01 CD4 and CD8 also serve as identity flags for different mature T cell subsets with different cellular functions.

3.3.6.9 According to the type of interactions between the developing cell and their environment, other T lymphocyte subsets develop.

3.3.6.9.1 Among the cell populations developing in the thymus, are regulatory CD4 T cells (Treg), a particular αβ T cell subset specialized in the control of the fate of other T cells.

3.3.6.9.1.1 Treg lineage choice being determined by the transcription factor Foxp3, upon selective environmental induction.

3.3.7 The final outcome of T cell differentiation is determined by the strength of TCR and co-receptor interactions and influenced by the factors present in the immediate environment of the developing cell, that may limit the number of cells in each population.

3.3.8 Ultimately, lymphocyte maturation depends on the coherence and the fine-tuning of signals collected by the developing cell.

3.3.8.01 Lymphocytes can only fulfill their development if they can find in the host resources to sustain their survival.

3.3.8.1 In the end, B and T cells can only exist and survive, if they can be coherently part of a "state".

◆◆◆

3.4 Immune-competent cells are continuously produced throughout life.

3.4.1 Each mature lymphocyte expresses a single type of diversified receptor with a unique recognition pattern.

3.4.2 Of two alleles coding for the chains of the antigen-specific receptors only one is expressed—allelic exclusion.

3.4.2.1 Since the number and density of engaged receptors may determine response or non-response, the expression of a single type of receptor ensures precise cellular responses.

3.4.2.1.01 This rule is not stringent and exceptions exist with cells expressing more than one receptor and conferring more adaptability to antigen-specific receptor repertoires.

3.4.2.1.02 Indeed, the rare dual expressing cells found in the host express quantities and combinations of the receptor chains that better fit their environmental conditions.

3.4.2.2 Lymphocytes that express an identical antigen-specific receptor belong to the same "clone".

3.4.2.3 The number of immune-competent cells (clones) present in the host sets the maximum "recognition" capacity of the IS at a given moment in time.

3.4.2.3.1 The cohort of different antigen receptors expressed by the existing immune-competent cells in that given moment defines the available repertoire.

3.4.2.4 However, the IS has the potential to recognize all possible shapes.

3.4.2.4.1 Thus, the universe of shapes that can be recognized by the available repertoire of the IS at any time point represents only a minor fraction of its potential recognition capacity.

3.4.2.5 The continuous production of new immune-competent cells confers to the IS the ability to fully exploit its potential diversity.

3.4.2.5.01 It allows renewing pre-existing cells with new cells expressing different receptors.

3.4.2.5.1 This approach permits the IS to evolve with time and better adapt to new different challenges.

◆◆◆

3.5 In all complex organisms the number of lymphocytes is limited.

◆◆◆

3.6 In these complex organisms there is a continuous daily production of new lymphocytes.

3.6.1 Hosts with reduced production of new lymphocytes still attain the numbers present in individuals with normal production rates.

3.6.2 The potential to produce new lymphocytes, either as the result of de novo production of new cells expressing a different set of receptors or as a result of the proliferation of pre-existing cells in response to environmental changes, by far exceeds the number required to replenish the host.

3.6.3 Thus, the number of either B or T cells is not limited by their rates of production.

3.6.3.1 It is either limited by the availability of trophic factors or "quorum-sensing" mechanisms.[11]

3.6.3.2 In contrast the number of immediate B and T cell precursors is directly determined by the number of available pluripotential progenitors.

3.6.3.2.1 Meaning that lymphocyte differentiation is a linear "conveyor belt" process, that once engaged proceeds stepwise to the end and that there are no defined specific niches to gather and control these progenitor populations.

3.6.4 Lymphocyte development is orchestrated by a probabilistic succession of events that require the intervention of multiple factors.

3.6.4.1 Absence of one of the involved factors leads to lack of lymphocyte development.

3.6.4.2 These genetic defects can be corrected by replacing the missing factors.

3.6.4.3 As the number of lymphocytes is not limited by their rates of production, it suffices to add a few corrected progenitors to replace the missing populations in the host.

3.6.4.4 The corrected progenitors must be able to exploit the unused environmental factors and out compete the defective hosts progenitors.

3.6.5 These properties provide a basis for cell therapy strategies.

◆◆◆

[11] Many species of bacteria use quorum-sensing to coordinate their gene expression according to the local density of their population. In 1965, W.S. Bullough proposed that soluble growth inhibitions, "chalones", produced by an organ would contribute to limit that organ size.

3.7 In an IS where new lymphocytes are continuously produced in excess but their number is kept constant, newly generated cells must compete with other newly produced or resident cells to survive.

3.7.1 This implies that lymphocyte survival is not a passive phenomenon, but rather a continuous active process driven by a highly selective environment.

♦♦♦

3.8 Competition can be defined as "an interaction between two populations leading to a reduction in the survivor, growth and/or reproduction of the competing individuals concerned".[12]

3.8.1 Competition is a consequence of the organization of the host environment and not an end in itself.

3.8.2 Mutual aid may offer a competitive advantage to some cells.[13]

♦♦♦

3.9.1 Competition may arise through different processes

3.9.1.1 Two populations interact directly with each other and one population can prevent the second population from occupying its habitat or from using the resources available—interference competition.

3.9.1.1.1 In this case competition is loosely related to resource level.

---

[12] After Begon, Harper & Townsend.
[13] "This view, however, I could not accept, because I was persuaded that to admit a pitiless inner war for life within each species, and to see in that war a condition of progress, was to admit something which not only had not yet been proved, but also lacked confirmation from direct observation". —P. Kropotkin

3.9.1.2 Populations may share the need for resources in limited supply—exploitation competition.

3.9.1.2.1 In this case competition is directly related to the level of resources available.

3.9.1.3 In general, competition is brought about by a shared requirement for a resource in limited supply.

◆◆◆

3.10.1 We define "resources" as "any factor, which can lead to increased cell survival or growth through at least some range of their availability"[14] or accessibility.

3.10.1.01 Resource is any factor, which is "used" by a cell and for that reason is no longer available to other cells.

3.10.1.02 Resource consumption by lymphocytes can therefore modify the fate of other cells.

3.10.2 Resources can be essential, complementary, substitutable, antagonistic or even inhibitory.

◆◆◆

3.11.1 Lymphocytes specialize in the use of their unique antigen recognition receptors to capture trophic survival resources.

3.11.1.01 Lymphocytes may nevertheless use multiple other resources (see below).

◆◆◆

---

[14] After Pianka.

3.12 Lymphocyte survival is thus, primarily determined by signals received by their antigen-specific receptors.

3.12.1 B cells without BCR have a poor survival probability.

3.12.2 T cells without TCR have a poor survival probability.

3.12.2.1 In hosts that lack the correct identity marker, the restricting MHC element, recognized by the TCR, most T cells fail to survive.

3.12.3 Recognition is, however, degenerate and each antigen specific receptor interacts with a complex shape space.

3.12.3.1 The assembly of ligands recognized by each receptor will define a unique shape space.

3.12.3.2 For B cells this is represented by the pool of recognized 3D molecular shapes.

3.12.3.3 For T cells it is represented by the collection of the host's identity markers and its modifications that can be recognized by the TCR.

◆◆◆

3.13.1 Ligand recognition by antigen-specific receptors may induce lymphocyte proliferation and survival.

3.13.1.1 Antigen is, therefore, a resource that can be competed for by lymphocytes.

3.13.1.2 However, in a lymphocyte population, as each cell expresses a different receptor, the multiple diverse receptors expressed by a cohort of cells of that population ensures recognition of a universe of shapes.

3.13.1.2.1 Therefore, at a population level antigen is never limiting.

3.13.1.2.2 That is to say that the number of lymphocytes cannot be limited by the availability of antigens.

3.13.1.3 Thus, if antigens are not in limited supply (at a population level) an antigen can only act as a resource because of the access it provides to some other resource, which is or may become limited in supply.

3.13.1.4 Antigens act as surrogate resources.

3.13.1.4.1 Antigen recognition induces cells to either express new trophic receptors or increase substrates that allow lymphocytes to use other resources in limited supply.

3.13.1.5 At the clonal level, antigen is a critical surrogate resource. That is to say, for lymphocytes that express a particular receptor there is a "limited" amount of specific antigen that is recognized by that receptor.

3.13.1.5.1 It determines the size of each clone depending on the extent of the shape space recognized by the antigen receptor expressed by the cell of the clone.

3.13.1.5.2 Limiting quantities of an antigen will limit the clone access to resources in limited supply.

3.13.1.5.3 Thus, lymphocytes with broader shape space of recognition—degenerate—have bigger clone size than lymphocytes with refined recognition patterns.

3.13.1.6 Resources include all factors, other than those recognized by the antigen-specific receptors, which, within a certain range of concentrations, favor lymphocyte survival or proliferation.

◆◆◆

3.14 Lymphocytes that cooperate in the host's economy, use and provide mutual resources, i.e., can be part of a "state" and are selected to survive.

◆◆◆

Chapter 4. Organization

4 Lymphocyte populations compete.

4.1 Competition occurs between cells of the same clone bearing the same antigen receptor—intra-clonal competition—or between cells of different clones with different antigen receptors—inter-clonal competition.

◆◆◆

4.2 The presence of competitors modifies the equilibrium size of the competing populations,

4.2.1 i.e., the numbers of cells from a lymphocyte clone is modified by the presence of other lymphocyte clones.

◆◆◆

4.3 The presence of competitor cells alters the population dynamics, i.e., the rate of growth and the life span of the competing cells.

4.3.1 Thus, lymphocyte life expectancy is not entirely an intrinsic property of the cell. It varies according to the environment and to the presence or absence of other competing cell populations.

◆◆◆

4.4 Population size can be modified by resource manipulation.

4.4.1 The number of lymphocytes specific for a particular antigen is modified by the quantities of that antigen. Antigen levels also modify the equilibrium between different lymphocyte populations.

4.4.1.01 During lymphocyte development, the presence of small amounts of antigen favors the survival and accumulation of the antigen-specific B cells.

4.4.1.02 In hosts with an artificially reduced number of cells bearing a particular MHC molecule, the number of T cells bearing receptors capable of recognizing these markers decreases.

4.4.1.03 This reduction in the number of cells expressing MHC results in competition between T cell clones recognizing the same MHC molecule.

4.4.1.04 However, MHC molecules are generally not in limited supply. MHC molecules recognized by the TCR acts as a surrogate resource allowing T cells to access other resources in limited supply.

◆◆◆

4.5 In ecological systems, the presence of competitors induces changes in morphology—character displacement—or niche shifts of the competing individuals.

4.5.1 Similarly, the characteristics of a lymphocyte population can be altered according to the presence or absence of other competing populations.

4.5.1.01 The presence of an abundant B cell clone alters the equilibrium and the relative representation among the remaining B cells clones.

◆◆◆

4.6 Lymphocyte survival relies on cell/ligand interactions, the availability of other "resources" and the nature and number of competing rivals.

4.6.1 Thus, in their continuous flight for survival lymphocytes must acquire a selective advantage over their competitors.

4.6.1.1 They must adapt to the immediate environment either by modifying their survival threshold requirements or through differentiation.

4.6.2 At different life stages, lymphocytes will require different signals to survive. These factors apply a continuous selective pressure throughout the lymphocyte life history.

◆◆◆

4.7 According to the "competition exclusion principle", lymphocyte competition should lead to the progressive dominance of a limited number of cell types (clones) and consequently to the exclusion of most clones.

4.7.1 Thus, in small closed environments, competition would be incompatible with maintenance of the immune system (IS) diversity—that is, with the presence of multiple lymphocyte clones able to recognize an universe of shapes.

4.7.1.01 Without diversity, the role of the IS collapses, as it would no longer be able to recognize the universe of "invader" shapes.

4.7.1.1 However, within larger areas of space, "differences in the age and structure of the populations, their migration capacity, the heterogeneity and patchwork distribution of fragment habitats and their change with time"[15] prevent the competition exclusion principle.

4.7.1.2 This explains the co-existence of multiple potentially competing species within larger areas of space. Cells home to environments where they compete best.

♦♦♦

4.8 The IS bypasses this "competition-diversity paradox" and selects preferentially for diversity through different mechanisms.

4.8.1 The continuous generation of cells ensures the permanent availability of new and very diverse clones, some of which may have the capacity to replace old cells.

4.8.1.1 Indeed, the progressive decrease in lymphocyte production observed with aging correlates with an increased frequency of pauciclonal repertoires and attests to its role in the IS diversity.

4.8.2 The control of cell proliferation by terminal cell differentiation followed by cell death also prevents unlimited clonal expansion and dominance.

4.8.2.01 In terminal differentiation, cells deviate all protein synthesis mechanism to the production of a single effector molecule and fail to ensure metabolic pathways needed for survival.

---

[15] After I. Hanski.

4.8.2.1 After resource usage lymphocytes down-regulate receptors "releasing" resources to be freely available to other populations.

4.8.2.2 These mechanisms facilitate the access of new cells, since established populations die or modify their resource requirements allowing the entry and establishment of new clones.

4.8.3 The capacity to use multiple alternative factors to survive and proliferate provides a critical advantage for the maintenance of cell diversity in a competing environment.

4.8.3.1 Each lymphocyte having its adequate "niche".

4.8.3.2 Niches cannot be defined by a single cellular property.

4.8.3.2.1 Niches can be represented as multi-dimensional spaces (n1, n2, n3, n4, n5, ...n∞) defined by the different resources that affect survival and proliferation of a particular cell type.

4.8.3.2.2 Thus, niches can be represented by a series of values (1, 47, 23, -7, 0, etc...) each corresponding to a particular dimension (resource value).

4.8.3.2.2.1 Substitutable sets of resources overlap in the same dimension.

4.8.3.2.3 The relative role of each resource being represented by the quantitative value attached.

4.8.3.2.3.1 A value of "0" implies that a particular resource is not required and its dimension is "curled" for that particular cell subset.

4.8.3.2.3.2 A negative value implies that a particular resource has a negative effect on the cell population studied.

4.8.3.2.4 Too much of a positive resource may also result in a counter-intuitive effect by increasing the effects of a negative resource within the niche.

4.8.3.2.4.1 Through this mechanism a favorable resource may, when in excess, become inhibitory.

4.8.3.2.4.2 This happens during lymphocyte development. Too strong interactions with some "resources" (antigen) can hinder, instead of favor, the development of the antigen-specific cells.

4.8.3.2.4.3 Moreover, during development, lymphocytes progress through different signal thresholds and at some steps are more sensitive to antigen encounters.

4.8.3.2.5 During differentiation, lymphocytes open access to "hidden" dimensions and become competent to use new resources.

4.8.3.3 The use of multiple resources by different lymphocyte populations allows the co-existence of distinct cell types.

4.8.3.3.1 For each cell line a hierarchy of resource relevance can be drawn.

4.8.3.3.2 A niche may be defined using only the most relevant resource information and suppressing details.

4.8.3.3.3 Levels of the most relevant resources may be modified thereby changing the niche size and the numbers of cells of particular subsets.

4.8.3.4 These parameters contribute to create a heterogeneity of habitats that favors co-existence of diverse potentially competing populations.

4.8.3.5 Thus, the variety of resources and the heterogeneity of habitats (in the secondary lymphoid tissues) allow different lymphocyte types to find an ecological niche to survive.

◆◆◆

4.9 Cells of multicellular organisms can only exist if they are part of a "state", thus a form of dependence.

4.9.1 Multicellular organisms congregate cohorts of cells which otherwise cannot survive on their own.

4.9.11 They are an example of mutualism among cell populations.

4.9.2 Lymphocytes also can only survive if they are part of a "state".

4.9.2.1 Mutualistic interactions shape lymphocyte populations.

4.9.2.2 Lymphocyte cooperation confers a competitive advantage to some cells and favors the maintenance of rare cell populations and the IS diversity.[16]

---

[16] "Who are the fittest: those who are continually at war with each other, or those who support one another?" P. Kropotkin.

4.9.2.3 It provides the necessary selective pressure for the survival of the "fitted" cells.

4.9.3 Cooperation and conflict interactions between lymphocytes equally contribute for the final composition and organization of the IS.

◆◆◆

4.10.1 During their development lymphocyte clones at different stages of differentiation modify either their resource requirements or interact differently with the same resources.

4.10.1.1 Cell differentiation is associated with the acquisition of receptors for different resources and growth factors,

4.10.1.2 and new migration and homing receptors.

4.10.2 Migration ensures the distribution of lymphocytes from original source to the host's tissues and between the different environments.

4.10.2.1 Migration allows the cell to find the appropriate niche to survive.

4.10.2.2 The inability of the cells to migrate into or find the appropriate niche compromises their survival.

4.10.3 The variety and the patchy distribution of resources in different habitats play a critical role in cell localization and differentiation.

4.10.4 Lymphocytes are highly motile cells that recirculate continuously in the body from blood to lymph and back, through secondary lymphoid structures.

4.10.5 Different lymphocyte populations populate different habitats.

4.10.5.1 Thus, B and T cells express different sets of adhesion and chemokine[17] receptors and therefore home to and occupy different areas in the lymphoid tissues.

4.10.5.1.01 Chemokines guide lymphocyte movement and migration. Lymphocytes migrate following chemokine gradients towards the source of the chemokine.

4.10.5.1.02 Chemokines play a role in directing lymphocytes to the lymph nodes or into other tissues and inflammatory sites.

4.10.5.1.03 Diverse sets of chemokines and receptors determine selective migration and homing.

4.10.5.1.04 Saturation of chemokine receptors leads to a refractory sate where the cell is desensitized and no longer responsive to that chemokine. It releases lymphocytes from a docking site.

4.10.5.1.05 Random cell motility—chemokinesis—and adhesion molecules further facilitate lymphocyte selective migration across tissues.

4.10.5.1.06 Lymphocyte movement may mimic "swarm intelligence" behavior, where the interactions between different individuals may lead to emergence of new global behaviors.[18]

4.10.5.1.07 These properties allow all lymphocytes to continuous circulate from blood to lymph and back to blood (too find the appropriate niche?).

---

[17] Chemokines are proteins with the ability to induce directed chemotaxis in nearby responsive cells. Chemotaxis is a phenomenon in which cells direct their movements according to certain chemicals in their environment.
[18] After Bonabeau, Dorigo and Theraulaz.

4.10.5.1.08 Lymphocytes leave the blood stream to enter the lymph nodes by migrating across specialized high endothelial venules (HEV) with an extraction rate of about 1 in 4 circulating cells. They exit lymph nodes to the lymph at different times after entry according to their subtype and state of activation.

4.10.5.1.09 Alymphopenic mice lack HEV. T cell transfer into these mice restores HEV structures.

4.10.5.1.1 B cells express receptors that direct them to home to the B cell follicles.

4.10.5.1.2 T cells occupy the periarteriolar and paracortical areas of the secondary lymphoid tissues.

4.10.6 The anatomical size of the niche may also play a determinant role in the control of cell populations.

4.10.6.1 This is the case for the B cell colonization of the secondary lymphoid tissue follicles.

4.10.6.2 In the small closed environment of the follicle, B cells are forced to compete for trophic factors present in limited supply, which limits the number of surviving B cells.

4.10.6.2.01 The presence of an invader may change the follicle environment e.g., through the release of inflammatory cytokines. Thus, it may increase resource availability and favor survival of expanded selected B cell populations.

4.10.7 Thus, the interplay between immune cells and their immediate environment determines control of cell numbers.

4.10.7.1 Lymphoid tissue disorganization may lead to uncontrolled lymphocyte proliferation or death.

4.10.8 Lymphocytes help to define their appropriate niches by releasing factors involved in tissue development.

4.10.8.1 They contribute to the organization of their own habitat.

4.10.8.2 Thus, the structure and organization of the secondary lymphoid organs are also determined by factors secreted by lymphocytes.

4.10.8.2.01 Development of lymph nodes is initiated early in ontogeny by specialized "inducer" cells of hematopoietic origin and "organizer" cells of stromal source. It is also modulated by lymphocyte seeding and colonization.

4.10.9 Lymphocytes produce factors (cytokines) that favor both their own survival and proliferation and that of other cell populations.

◆◆◆

4.11.1 Different lymphocyte populations use different resources.

4.11.2 The numbers of T and B lymphocytes are independently regulated.

4.11.2.1 The number of B cells is stable in the absence or in the presence of T cells; T cells numbers remain similar in the absence or in the presence of B cells.

4.11.2.1.1 However, when both populations are present they interact and mutually influence each other.

4.11.2.2 T and B cells belong to different guilds as they use different recognition strategies and exploit different resources.

4.11.2.3 B cells survival implies BCR engagement,

4.11.2.3.1 while T cell survive through the TCR recognition of the MHC host markers.

4.11.3 B cell express receptors that specifically recognize a B-cell activating factor called BAFF.

4.11.3.1 Changes in the levels of BAFF modify B cell numbers.

4.11.311 Thus, it can be said that B cells exploit BAFF.

4.11.3.1.2 It is likely that BCR signals modify the B cell access and use of BAFF as a survival resource in limiting supply.

4.11.4 Among T cells, the number of CD4 and CD8 T cells is co-regulated.

4.11.4.01 In the absence of CD4 T cells there are more CD8 T cells; in the absence of CD8 T cells there are more CD4 T cells.

4.11.4.1 The numbers of CD4 and CD8 T cells are interrelated indicating that they partially share common resources.

4.11.4.2 CD4 and CD8 T cells recognize primarily different host identity markers.

4.11.4.2.1 Thus, the resources shared by CD4 and CD8 T cells cannot be only related to antigen-specific recognition.

4.11.5 T cells exploit mainly cytokines—interleukins 2 (IL-2), 7 (IL-7), 15 (IL-15) as survival factors.

4.11.5.01 TCR signals modify the ability of the T cells to use these cytokines and other "nutrients" as survival resources.

4.11.5.1 T cells express different combinations of cytokine receptors,

4.11.5.1.1 and changes in the levels of these interleukins can alter the number of different T cell subsets.

◆◆◆

4.12 B and T lymphocytes express different antigen specific receptors that recognize an infinite number of shapes and once engaged signal lymphocytes to respond.

4.12.01 Without these signals lymphocytes cannot be part of a "state" and cannot survive.

4.12.1 However, the strength of ligand binding, receptor occupancy and cross-linking, the engagement of additional co-receptors and the resource level in the surrounding environment will also determine lymphocyte fate.

4.12.2 Thus, upon receptor engagement lymphocytes can either (i) survive, (ii) get active, (iii) get active and divide, (iv) get refractory to new stimuli or (v) die.

◆◆◆

4.13.1 Within the body's economy and considering the vast diversity of BCR and TCR expressed, there is a high probability that some B and T cell will find the correct interactions to be activated.

4.13.2 Upon activation, lymphocytes engage in niche differentiation and acquire new survival requirements thereby avoiding competition with non-activated cells.

4.13.2.1 They adapt to new niches.

4.13.2.2 Thus, after activation, cells of the same clone differentiate and either interact differently with the same resources or modify resource requirements,

4.13.2.2.1 It is as if they open curled dimensions to add new valid numerical values to their evolving definition of niche.

4.13.2.3 Lymphocytes in different states or with different histories have different signal thresholds.

4.13.2.4 Lymphocytes achieve this by tuning their responses to different patterns of signals.[19]

4.13.3 The IS is structured to host both resting (naïve) and activated populations of lymphocytes

4.13.3.01 Naïve cells are cells that have completed maturation, but have not interacted in a "significant manner" to be permanently modified by their specific-receptor ligands.

4.13.3.02 Activated/"memory" cells are cells that have been modified upon interaction with antigen.

4.13.3.1 The populations of activated and naïve cells show different resource requirements.

4.13.3.2 The frontiers between niches of the two cell populations being defined by progressive changes in survival requirements during the transition from the naïve to the activated/"memory" state.

◆◆◆

4.14 In order to fulfill its tasks, the IS requires a variety of lymphocyte subpopulations that play specific roles.

---

[19] After Z. Grossman and W. Paul.

4.14.1 The homeostatic mechanisms that control lymphocyte numbers must preserve different minority subpopulations and maintain their relative equilibrium.

4.14.1.1 Both absolute numbers in each sub-population and their relative sizes must be maintained.

4.14.2 In the IS the different cell populations stand in a determined relation to one another.

4.14.2.01 Example 1: B and T cell belong to different guilds and occupy different habitats,

4.14.2.02 Example 2: CD4 and CD8 T cells partially share resources and habitats.

4.14.2.1 The way in which different cell populations are connected is the structure of the IS. There are several types of interrelation between the diverse cellular subpopulations of the IS.

♦♦♦

4.15 The IS maintains populations of resting and activated B cells.

4.15.1 B cell populations show a pattern of V_h-gene expression that is highly conserved and tightly regulated.

4.15.1.1 This observation suggests that BCR interactions that lead to B cell survival do not require the involvement of the full antigen-binding site.

4.15.1.2 Thus, interactions of lower strength, low "avidity",[20] may suffice to promote B cell survival.

---

[20] Avidity is a term used to describe the combined strength of multiple bond interactions. Avidity is distinct from affinity, which is a term used to describe the strength of a single bond.

4.15.2 Stronger, high avidity BCR interactions will lead to B cell activation, proliferation and further differentiation that culminate in Ig production and secretion.

4.15.2.1 We have discussed that in the body's economy, considering the vast diversity of BCR expressed, there is a high probability that some B cells will find the correct interactions to be activated.

4.15.2.2 Thus, in each host a fraction of the B cells is activated, differentiates and secretes Igs.

4.15.2.2.01 In absence of T lymphocytes, B cells secrete mainly an Ig class in a pentameric form—IgM.

4.15.3 The numbers of non-activated (naïve) and activated Ig-secreting B cells are independently regulated.

4.15.3.1 The lack of correlation between the number of B cells and the number of Ig-secreting cells indicates that these two cell populations have different resource requirements.

4.15.3.1.1 Indeed, in presence of very low numbers of B cells, a significant fraction of these cells is activated and the number of IgM-secreting cells and the IgM levels in the serum of the host are normal.

4.15.3.2 Serum IgMs are generally produced by particular subsets of B cells, namely B cells that occupy the marginal zone (MZ) around the follicles in the lymphoid tissues and B1 cells.

4.15.3.2.1 When these MZB or B1 cells are missing, other B cells types can differentiate and take the relay to reestablish normal quantities of serum IgM.

4.15.4 Once the number of activated IgM-secreting B cells reaches equilibrium, naïve B cells accumulate.

4.15.4.1 This implies the existence of a hierarchical organization of the IS.

4.15.5 The two B cell populations are connected in a hierarchical way in which the replenishment of the activated B cell compartment is a priority.

4.15.5.01 Indeed, secreted IgMs bind to a broader panel of antigens and help to eliminate bacterial invaders. They are part of the innate IS.

4.15.5.1 Thus, the IS is structured to first ensure the maintenance of normal levels of natural IgM antibodies, which constitute an initial barrier of protection against bacterial infections,

4.15.5.2 and keeps a reserve of naïve B cells to allow more efficient "adaptive" responses to new challenges.

4.15.5.3 The population of activated B cells self-renews through cell division and thus can persist autonomously for relative long periods of time.

4.15.6 The number of activated B cells is strictly controlled.

4.15.6.1 Once established, activated B cells prevent new incoming B cells to be activated by feedback mechanisms involving Ig secretion.

4.15.6.1.1 This implies that B cells may use Ig levels as quorum-sensing mechanisms that limit the number of activated B cells.

4.15.6.1.2 Indeed, B cells express inhibitory receptors that interact with the constant part of Ig (FcR) and hinder B cell activation.

4.15.6.2 Replenishment of the compartment of activated B cells follow the rule "first come, first served".

4.15.6.2.01 As consequence it is enriched in B cells formed early in ontogeny with a broader pattern of recognition, as they were selected in the presence of an excess of resources.

4.15.6.2.1 This, however, is not fail safe since activated Ig-secreting cells are in most instances terminally differentiated plasma cells,

4.15.6.2.1.1 i.e., they become so specialized that they only produce Igs and no longer other house keeping products necessary for survival.

4.15.6.3 Replacement of these cells occurs when newly formed B cells show higher competitive fitness to get activated and survive.

4.15.6.3.1 Replacement is therefore related to the diversity of the BCRs expressed by newly formed B cells and the degree of specialization of their recognition space

4.15.6.3.2 and the antigenic environment to which the host is exposed.

◆◆◆

4.16.1 The IS accommodates populations of naïve and activated T cells.

4.16.2 The naïve and the activated CD8 T cell populations are connected in a way in which the two compartments (naïve and activated) are regulated by independent homeostatic mechanisms.

4.16.2.1 The number of naïve cells is partially dependent on the production of new cell types in the thymus and is not generally affected by the number of activated cells.

4.16.2.2 Activated cells can exist in "normal" numbers in the absence of naïve cells.

4.16.3 CD8 T cells require TCR signals to survive,

4.16.4 and it has a probability "p" of encountering the correct TCR ligands and environmental conditions that favor activation.

4.16.4.1 This probability increases for CD8 T cells with broader shape space of recognition.

4.16.4.1.1 The pool of activated cells is therefore enriched with cells with degenerate patterns of recognition.

4.16.4.1.2 This may allow the cohort of activated cells to cover a broad shape space territory in highly competitive conditions that would otherwise lead to the establishment of clonal dominance and restricted recognition repertoires.

4.16.5 After activation, CD8 T cells tune their threshold of survival in response to recurring signals,

4.16.5.1 modify their response to cytokines and acquire the capacity to exploit new resources.

4.16.5.2 The modified response of cells after activation contributes to the independent homeostatic controls of naïve and activated CD8 T cells.

4.16.6 The independent control of the naïve and activated CD8 T cells allows the IS to use the populations of activated cells to respond rapidly to existing environmental invaders and at the same time to maintain a reservoir of diversity to deal with new challenges.

◆◆◆

4.17 The IS must be also organized to maintain minor lymphocyte populations.

4.17.1 A specific cell has a higher probability of survival if it possesses superior competitive fitness, that is, a superior ability to exploit resources common to competitors,

4.17.1.1 or a unique ability to specialize, thereby allowing the exploitation of alternative resources.

4.17.2 Thus, maintenance of a minority population of cells depends on whether specialization provides to the cells a selective advantage in the utilization of a unique trophic resource.

4.17.3 A subset of CD4 T cells exerts regulatory functions and can inhibit the expansion of other CD4 T cells subsets.

4.17.3.1 These Treg cells produce factors that interfere with survival, expansion or function of other lymphocyte types.

4.17.4 This population of Treg cells is stable.

4.17.4.1 Treg cells are characterized by the expression of high affinity chain of the receptors for IL-2 (IL-2Rα).

4.17.4.2 Parsimony tells us therefore that they are highly specialized for the exploitation of the IL-2 resource.

4.17.4.2.1 They occupy a niche defined by the IL-2 resource, which allows them to avoid direct competition with other T cells lacking the high affinity IL-2Rα chain.

4.17.4.2.2 This specialization explains how the IS ensures the presence of the minor sub-population of Treg cells.

4.17.4.3 The size of the Treg niche corresponds to available quantities of IL-2.

4.17.4.4 However, the situation is more complex than simple selection by direct competition and the occupation of a specialized niche.

4.17.4.4.1 Other CD4 T cells make IL-2.

4.17.4.4.2 Therefore the homeostasis of the Treg cells is also (and mainly) dependent on the IL-2 produced by other CD4 T cell populations.

4.17.4.4.3 The number of Tregs is tied, or indexed, to the number of CD4$^+$ T cells that can make IL-2.

4.17.4.5 The indexing of Treg cells to the CD4 T cells that may produce IL-2 creates an interdependence; increases or decreases in numbers of the latter will be reflected in the expansion or contraction of the former, thereby maintaining the relative proportions.

4.17.4.5.1 Since Treg cells limit the proliferation of other CD4 T cells; this interdependence constitutes a feedback mechanism to control T cell numbers.

4.17.4.5.2 In this case the two CD4 populations are connected in an interdependent way that explains why the relative proportion of the two populations is kept stable.

4.17.5 This implies that IL-2 may act as part of quorum-sensing mechanisms that limit the number of CD4 T cells.

4.17.6 Independent populations keep relative proportions provided that the levels of the specific resources that they use is kept constant.

◆◆◆

4.18 Lymphocyte populations cooperate.

4.18.1 Through cooperation the survival and growth rate of one lymphocyte population is increased in the presence of a second population,

4.18.1.1 the numbers of cells from a proliferating clone is increased by the presence of other proliferating lymphocyte clones.

4.18.2 Lymphocytes produce resources that (i) shape cell environment, (ii) facilitate (help) survival and proliferation of other cell types.

4.18.3 Subsets of CD4 T lymphocytes specialize in these helper functions.

4.18.3.01 As T cells can produce different cytokines, there are as many CD4 cell subsets as cytokines a CD4 T cell can make.

4.18.4 In steady state, the relative proportions of different lymphocyte populations is in part dictated by these helper effects.

4.18.4.1 CD4 T cells that make IL-2 help Treg cells and other cell populations that express receptors for IL-2 (activated CD8 or CD4 T cells).

4.18.4.2 CD4 T cells that make IL-4 help B cell to survive and promote their differentiation to produce different Ig (IgG) isotypes.

4.18.4.2.1 This implies the activation of the B cells specific enzymes responsible for class switch Ig recombination (AID).

4.18.5 Through cooperative interactions lymphocyte populations can co-evolve in the host.

4.18.5.1 Thus, CD4 T cells can select B cell sub-sets that otherwise are absent from the body, conversely the relative representation of distinct T cell subsets may be altered in the presence of B cells.

4.18.6 Helper resources may act independently of the type of receptor ligands involved—bystander effects—or imply antigen specific cell-to-cell cooperation.

4.18.6.1 They will determine the number of responding B and T cells and their final differentiation into the activated effector/"memory" compartments.

4.18.6.2 Thus, increased resources produced or induced by the activated CD4 T cells are critical for the development of immune responses.

◆◆◆

4.19 CD8 T lymphocytes that specialize in killing target cells can kill other lymphocytes and modify the relative equilibrium between populations.

4.19.1 Thus, predator-prey interactions also shape lymphocyte populations.

4.19.1.01 CD8 T cells can, for example, eliminate transformed tumor lymphocytes.

4.19.1.001 Elimination of host tumor cells, which express modified levels of MHC, is generally done by the natural killer (NK) cells part of the host's innate defenses.

◆◆◆

4.20 The way that the different lymphocyte populations are connected changes with time.

4.20.1 Early in ontogeny, during the host's growth, lymphocyte populations expand to reach an equilibrium state.

4.20.2 In the expansion phase, the excess of growth factors and resources allow the IS to accommodate all newly formed cells.

4.20.2.1 At equilibrium, lymphocytes reach the host's carrying capacity: that is the number of cells that the host environment can support.

4.20.2.2 Resources become limiting and any newly generated cell can only survive if another resident cell dies.

4.20.2.2.01 Alternatively expanding lymphocyte populations may release factors that limit further cell accumulation through a quorum sensing mechanism.

4.20.3 These dynamic changes are accompanied by changes in lymphocyte repertoires.

4.20.3.1 Neonatal lymphocyte repertoires differ from those of adult hosts.

4.20.3.2 In neonates the vast majority of the antibodies are multi-reactive while in adults most of the antibodies are mono-specific.

4.20.3.3 These changes of cell repertoires observed in the transition from neonatal to adulthood correlate with overall modifications of cell population dynamics.

4.20.3.3.1 In neonates during the phase of expansion of lymphocyte numbers, cell selection is permissive, while in adult at equilibrium when resources become scarce and competition is established, selection becomes strict.

4.20.3.3.2 Thus, in adults, resource competition favors selection of cells with specialized and unique niche requirements.

4.20.4 The replacement of neonatal multi-reactive by mono-reactive repertoires suggests that during onto-geny, lymphocyte populations go through a process of "immunological succession" in which generalist cells are replaced by specialists.

4.20.5 Competition based on the receptor diversity and environment heterogeneity leads to the substitution of the initially selected population by new clones.

4.20.5.1 Substitution is thus determined both by the rate of colonization (invasion) by new cells and the rate of extinction of pre-established cells.

4.20.5.2 If a cell migrates into a niche that is already colonized the newly arriving cell either out-competes and displaces the established cell or is prevented from entering the niche and dies—pre-emptive competition.

4.20.6 The continuous daily production of new immune cells allows at least partial substitution of the established populations.

4.20.7 When both the new and the established cell express the same antigen receptor, substitution will depend on the state of the cells.

4.20.7.1 If both cells share the same state and resource requirements substitution occurs at random,

4.20.7.1.1 meaning that lymphocytes do not have an unique pre-fixed life expectancy.

4.20.7.2 A population pre-established in the correct niche and with a low extinction rate benefits from the founder advantage "first come, first served". It resists replacement by a new incoming cell and substitution does not take place.

4.20.7.2.1 However, the established cell may have been modified to exploit new resources, drifted to a different niche and in this case both cells co-exist.

4.20.7.3 If the established cell population has been "negatively" modified ("anergized"), substitution will take place and the new cells will replace the old cells.

4.20.8 The probability that two progenitor cells will rearrange exactly the same receptor is extremely low.

4.20.8.1 Thus, in general, the specificity and degeneracy of the antigen receptors expressed by the newly formed and the established cells will determine whether substitution takes place or not.

4.20.9 The presence of an early population may also modify the environment to facilitate colonization by a latter cell type.

◆◆◆

4.21.1 In a closed or full environment in presence of multiple competitors, the richness of the niche defines clone size.

4.21.1.1 The size of the niche, that is the numerical values of its dimensions, is the limiting factor that will determine the final number of surviving cells.

4.21.1.2 In this case, the number of arriving cells from the same clone is irrelevant, as only a limited number will survive.

4.21.1.2.1 If "x", "10x" or "100x" new identical cells arrive, only a fraction of "x" (the carrying capacity of each clone) will survive.

4.21.2 In an open or empty environment, without competitors, arrival of new cells may result in their expansion and accumulation if the cells can use the available unused resources.

4.21.2.01 This is partially the case during the neonatal expansion of the IS.

4.21.2.1 In hosts where lymphocytes lack particular receptors, the resources usually sequestered by these cells will be freely available.

4.21.2.2 The expansion of the new cells will be limited by the size of the exploitable niche—the availability of the free resources.

4.21.2.2.1 In this case, if "x/10", "x", "10x" or "100x" new identical cells arrive they will expand (or not) to reach the "y" number (the carrying capacity) but no more.

4.21.2.3 The specificity and degeneracy of the antigen receptor, the quantity and multiplicity of recognized antigens, and the access they provide to other resources determine the final clone size.

4.21.2.3.1 The dynamics of terminal differentiation and subsequent death is a determinant factor that explains the smaller size of B cell populations in adoptive lymphopenic hosts as compared to T cell populations.

4.21.2.4 The arrival of new cells may also induce the production of more resources that increase survival and proliferation of the new cells.

◆◆◆

4.22 Homeostasis of the IS is a highly dynamic process.

4.22.01 Thus, homeostatic equilibrium of different lymphocyte populations and their respective interconnections can be reached at different levels according to environmental cues.

4.22.02 Steady states will differ from individual to individual and in each individual over time and subject to the history record.

4.22.1 The IS will react to environmental changes to maintain its internal balance.

4.22.2 The initial steady state will determine the IS reaction to the presence of invaders.

◆◆◆

Chapter 5. Response and Memory

5. Interactions between host and alien invaders may have different outcomes: mutual adaptation and co-existence, death of the host or elimination of the invader.

5.01 From its point of view, the host must either avoid or keep the invader under control.

5.01.01 In extreme cases, in the presence of a viral invader, some eukariotes simply "disappear" i.e., they change from a diploid to a different haploid form with a different phenotype not recognized by the invader, the "Cheshire Cat" escape strategy.[21]

5.01.1 A role for the IS is to help the host to control or eliminate the invader.

5.02 From the point of view of the invader only the mutual adaptation alternative is valid in the long term.

5.02.1 Invaders prefer to establish stable symbiotic relations. Thus, many invaders have evolved multiple different strategies to prevent the intervention of the host IS.

5.02.2 Some create diversion and avoid specific and efficient responses by stimulating in a non-specific manner all cells of the IS.

5.02.3 Some avoid the IS by creating escape mutants that cannot be "recognized" and eliminated by ongoing responses.

5.02.4 Others selectively suppress the effector arms of the IS and avoid eradication.

[21] From Miguel Frada et al (PNAS, 2008).

5.02.5 Or simply destroy the host's IS, as is the case for the human immunodeficiency virus (HIV).

◆◆◆

5.1. In general invaders represent a perturbation of an established steady state,

5.1.01 that provokes an immune response (IR) by the IS.

5.1.02 Small or smooth non-invasive changes in environmental antigens may not suffice to provoke a reaction.

5.1.1 Once the perturbation ceases the IS returns to a new steady state.

◆◆◆

5.2.1 Invaders are complex associations of multiple structures.

5.2.2.1 Some of these structures (carbohydrates, proteins) interact directly with lymphocyte (B cell) receptors.

5.2.2.2 Others interact with innate defenses, reminiscent of primordial mechanisms that preserve the individuality of single cell protists,

5.2.2.2.1 resulting in the death and processing of the invader's structures.

5.2.3 In this process, fragments of the invader proteins (peptides) will be present in some host's cells where they will interact with other cells of the IS.

5.2.3.01 These cells act as antigen-presenting cells (APCs).

◆◆◆

5.2.3.1 APCs associate the invader's structures with MHC molecules that interact with (are recognized by) TCRs.

5.2.3.1.1 Any host cell may act as an APC depending on their capacity to process, associate and present peptides of the invader with the MHC.

5.2.3.1.2 However, each APC defines a particular and unique microenvironment that will determine the progression of the host's IR.

5.2.3.2 Cells of the reticulo-endothelial system (RES), monocytes, macrophages and dendritic cells (DCs) of various origins, are distributed through out the organisms and act as APCs at different states of differentiation.

5.2.3.3 Once modified by the invader, some of these cells mature and migrate from the local of invasion to the nearest secondary lymphoid organ.

◆◆◆

5.3 Lymphoid tissues provide favorable environments where the different actors of the IR, i.e., APCs and lymphocytes, meet and where resources concentrate.

5.3.1 IRs are highly dynamic processes that imply cell movement and migration.

5.3.1.1 Cells migrate to sites where they may interact with other cell types and gain a selective advantage to proliferate and survive.

5.3.1.2 Different cell types participating in an immune reaction may also meet to develop new organ structures—tertiary lymphoid tissues.

5.3.1.2.1 These structures are created during IRs in species, i.e., birds, that do not have peripheral secondary lymphoid tissues, such as lymph nodes.

◆◆◆

5.4 In an IS that maintains a highly diverse population of lymphocytes and is capable of recognize multiple antigens, the number of cells that can interact with the determinants of one particular invader is low.

5.4.1 The response of each individual lymphocyte is a probability function that will be determined by the levels of binding and cross-linking of the antigen-specific receptors and the engagement of additional co-receptors.

5.4.2 Lymphocytes that tune the correct level of signals have a higher probability of getting activated.

5.4.3 Upon activation selected genes are activated or silenced randomly, which implies that even within a clone of cells bearing the same antigen receptors each cell may follow a different outcome.

5.4.3.1 However, protection of the host cannot be achieved by the response of an isolated single cell.

5.4.4 Control of the alien invader requires the increased clone size of the specific lymphocytes.

5.4.4.1 Upon activation, lymphocyte clones that interact with the antigenic determinants of the invader proliferate and expand.

5.4.4.2 Thus, initiation of an IR is conditioned by the increase in availability of resources that favor lymphocyte proliferation and expansion.

5.4.4.2.1 Antigen represents the specific resource that triggers accessibility of the responding cells to other non-specific resources generally in limiting supply.

5.4.4.2.2 Local inflammation determined by the invader and its interactions with host's cells and tissues favors the increased production of non-specific resources that can be used by the responding lymphocytes.

5.4.5 IR are multifactorial probabilities determined by a set of interactions rather than a simple cohort of cells.

5.4.5.1 IR are influenced by the number of invaders, their rate of division and dissemination, the route of invasion, the type of the local APCs, etc.

5.4.5.2 The variety and the distribution of resources in different habitats will change the lymphocyte environment and ensure heterogeneity of the responses.

5.4.5.3 Engagement of the antigen specific receptors confers a selective advantage to antigen-specific cells by increasing their resource usage competence.

5.4.5.3.1 It determines the specificity of the response.

5.4.5.4 Progression of an immune response will also depend on the initial context—all aspects of past history relevant for future probabilities.

5.4.5.4.1 The initial microenvironment context in which invaders interact with the IS determines the type of IR.

5.4.5.4.2 There are IR that favor production of antibodies (humoral responses),

5.4.5.4.3 responses that favor inflammatory reactions initiated by T cells and their products (cytokines),

5.4.5.4.4 and others that favor killing of infected cells (cytotoxic).

5.4.5.4.5 These types of responses are not mutually exclusive and can co-exist either at different sites or at different times during responses.

5.4.5.5 The type of response will be determined by the history record of the IS previous experiences.

◆◆◆

5.5.1 Once an IR is initiated, antigenic oscillations can arise as a consequence of the increased number of antigen-specific cells and of dynamics of the responses to diverse determinants, which may be more or less represented.

5.5.1.1 Thus, there is a probability that, among the diverse clones responding to the multiple determinants of the invaders, some interact with higher "avidity" or are more represented or proliferate faster.

5.5.1.2 Which makes that of the overall responding clones only a few will prevail.

5.5.2 These processes determine the presence of dominant responses to single determinants.

5.5.3 After the initial phase of expansion, resource and antigen levels drop and the responding clones undergo a phase of contraction and only a fraction of the antigen-specific cells persist.

5.5.3.01 That is, once the perturbation ceases the cell populations of the IS return to equilibrium.

5.5.3.1 In the presence of diverse competing clones the cells persisting after the peak of the response will occupy a specific niche.

5.5.3.2 A dominant clone, which may have expanded 100- or 1000-fold to the peak of the response, will contract to a constant "y" number of cells—the carrying capacity of its environmental niche.

5.5.3.2.01 If the size of the niches is not altered, higher expansions may be followed by stronger contractions.

5.5.3.2.1 A non-dominant clone, which expands only 10-fold, will probably have to contract less to reach its carrying capacity.

5.5.3.3 Dominant and non-dominant clones may partially share available niches.

5.5.3.4 The variety of resources and the heterogeneity of niches allow co-existence of diverse dominant and non-dominant clonotypes.

5.5.4 Cells persisting after the contraction phase have engaged in niche differentiation and tuned their threshold signals to survive in less favorable conditions. They are the "fitted".

5.5.5 The IS adapts to a new qualitatively different steady state.

5.5.5.01 A state that now considers elements of past history that determine new possibilities.

5.5.5.1 Thus, upon a new contact with the same invader the host can mount a faster and more efficient secondary IR.

5.5.5.2 This is a memory response.

5.5.5.3 Memory responses can be elicited a long time after the clearing of the original infection.

5.5.5.4 Memory responses may also be strictly related to the original site of infection—memory of the local tissue (niche) of interaction.

5.5.5.4.01 T cells activated in the skin will tend to return to the skin upon challenge, while those T cells that are activated in the gut will tend to return to the gut.

5.5.5.4.02 This implies that activated cells differentiate to acquire new sets of adhesion and homing receptors according to their site of activation.

5.5.5.4.03 The ability to home to sites of invasion, provides the IS new sites to accommodate activated/memory cells.

5.5.6 Immune memory represents the establishment of a new kinetic state in the IS, which implies that responses to new invaders will be altered.

5.5.6.01 Past elements will impinge new properties and determine different responses to new antigenic challenges.

5.5.7 Memory responses are the basis of vaccination protocols to protect populations against pathogen invaders.

◆◆◆

5.6 B cells interact directly with antigens of the invader to produce antigen-specific antibodies.

5.6.1 In this type of IR the antibodies produced are mainly of the IgM isotype.

5.6.2 The amplitude and duration of this response vary as a function of the number of B cells engaged in the response, the rate of division and elimination of the invader.

5.6.2.1 During the initiation of the IR, elimination of the invader decreases the quantities of available antigen that leads to a subsequent decrease in the production and circulating levels of IgM antibodies.

5.6.2.1.1 IgM antibodies, however, have a short half-life in circulation and cannot reach all body tissues.

5.6.2.2 If elimination of antigen is not complete it determines a new wave of production of IgM antibodies to give rise to an oscillating response that resembles predator-prey type of interactions.

5.6.3 These patterns of antibody responses are characteristic of thymus-independent responses; they occur in the absence of T cells, meaning that naturally T cells do not participate in these responses.

5.6.3.01 In the absence of T cells, B cell responses cannot undergo efficient affinity maturation to produce better and more competent antibodies.

5.6.3.1 The thymus-independent responses are mainly IgM antibodies directed to polysaccharide determinants of bacterial surfaces.

5.6.3.2 Thus, IgM plays a critical role in the host's protection against bacterial infections.

5.6.4 Most antibody responses are thymus-dependent; they are amplified and modified by cooperative interactions of B cells with helper CD4 T cells.

5.6.4.1 Thus, at the same time as B cells interact directly with determinants of the invader, T cells interact with peptides from the same antigenic structure.

5.6.4.2 CD4 T cells interact either with antigen presented by MHC molecules on either the B cell or an APC in close vicinity, get activated and produce factors that favor B cell proliferation and survival.

5.6.4.2.1 These collaborative relations involve chemokines and chemokine receptors that determine mutual attraction of B and CD4 T cells to the interaction sites.

5.6.4.3 Helper CD4 T cells produce cytokines, e.g., IL-4, that promote Ig isotype switch and somatic hypermutation (see **5.7.2**).

5.6.5 During the primary T cell dependent IR, specific B cells proliferate and expand and some differentiate into terminal IgM antibody producing cells that help to control the invader.

5.6.5.1 In the presence of help T cell factors, some progeny of the B cells from the original responding clone, start producing switched Ig isotypes.

5.6.5.1.01 Because of their degree of specialization, the Ig-secreting plasma cells can only produce antibodies, are not able to ensure their own survival and most die soon after differentiation.

5.6.5.2 Once the invasion is under control, the diminished antigen levels and resources can no longer sustain expansion (perturbation ceases).

5.6.5.2.1 The number of antigen-specific cells enters a contraction phase after which only a fraction of the antigen-specific cells persist.

5.6.5.3 In presence of diminished antigen levels, B cells whose BCRs have higher affinity for the antigen will out compete other B cells and persist.

5.6.5.4 Selection of the high affinity B cells occurs in the germinal centers.

◆◆◆

5.7 Germinal centers develop in the B cell follicles during T cell dependent antibody responses and are one of the best-characterized ecological niches in the IS.

5.7.1 B cells that give rise to germinal centers are initially activated outside follicles and on the average three B cells colonize each follicle.

5.7.1.1 Clonal expansion of the initial founder cells, driven by antigen held by follicular dendritic cells and T cell help, prevents colonization of the germinal center by B cells specific for a second unrelated antigen.

5.7.2 Expansion of the IgM antigen-specific B cells is later accompanied by Ig isotype switch and BCR hypermutation, two processes that require T cell help.

5.7.2.1 These processes are mediated by the Activation Induced Deaminase (AID) enzyme, which is selectively expressed by activated B cells in presence of T cell help in germinal centers.

5.7.2.2 AID targets proteins required for switching, including DNA cleavage and repair enzymes, to specific switch regions of the IgH-chain loci.

5.7.2.2.1 Ig switch allows the B cell to produce antibodies that maintain the same "recognition" variable site, but with different effector functions.

5.7.2.2.01 The initial context and the environment where B cell activation occurs determines the type of isotype switch.

5.7.2.2.02 Switch to different IgG subtypes occurs during most types of peripheral or systemic infections, IgA switch occurs at mucosal sites and IgE switch upon parasite invasion.

5.7.2.2.03 IgA secretion at mucosal sites prevents invaders entry.

5.7.2.2.03.1 IgA secretion in the milk allows to the neonate its immediate and correct insertion into the mother's environment.

5.7.2.2.03.2 It provides a simple mechanism for the transfer of the memory of past microbial encounters in the immediate environment of the neonate.[22]

---

[22] That R. Zinkernagel refers as "herd immunity".

5.7.2.2.04 IgE secretion favors parasite eviction by increasing peristaltic movements.

5.7.2.2.04.1 This is achieved by recruitment of cells, which express receptors for the constant fraction of IgE and that, upon cross-linking of these receptors by antigen, release histamine.

5.7.2.2.04.2 Expression of IgE in inappropriate sites can lead to unwanted hosts reactions; e.g., asthma, allergy.

5.7.2.3 Switched antibodies are bivalent molecules and are generally more efficient than pentavalent IgMs due to their increased avidity for antigen and to improved accessibility to body tissues.

5.7.3 AID selectively modifies DNA sequences coding for Ig V-regions, changes that are set after recruitment of DNA repair mechanisms.

5.7.3.1 These mechanisms increase, at selective sites, the rate of mutation from a background of 10^{-9} to 10^{-3}/base pair/cell division.

5.7.4 Mutations in the Ig V-regions allow the B cell to express receptors with a higher affinity for antigen.

5.7.5 In the close environment of the germinal center, competition among the proliferating B cells based on their ability to interact with limiting antigen held on follicular dendritic cells (FDC), will lead to the preferential survival of the B cells with mutated receptors with higher affinity for antigen.

5.7.5.1 "High affinity B cells" can more readily capture BCR signals and exploit the available resources.

5.7.5.2 B cells whose BCRs have lost the ability to interact with the antigen following mutation are out-competed and lost.

5.7.6 Germinal centers therefore play a critical role in the maturation of IR by selecting cells with high avidity for antigen binding.

5.7.6.1 They seem to have evolved to provide the appropriate niche for selecting B cells that gain competitive advantage by somatic hypermutation.

5.7.6.2 Indeed, somatic hypermutation is present in phylogeny well before the development of the affinity maturation of IR.

5.7.6.3 Selection of B cells expressing high affinity receptors e.g., affinity maturation of the IR, however, is only present in species capable of germinal center formation.

◆◆◆

5.8 Upon challenge with the same invader, the IS mounts a secondary (recall) antibody response that is faster and more efficient than the primary IR.

5.8.1 The antigen-specific B cells persisting after affinity maturation express switched BCRs with high affinity for the antigen—they are "effector/memory" B cells.

5.8.1.01 Activated IgM expressing B cells, which were shown to persist in time may represent a reservoir of true memory B cells ready to switch and enter the effector/memory stage.

5.8.1.1 Memory B cells show a faster rate of cell division and after differentiation produce more efficient switched antibodies.

5.8.1.2 Secondary IR can be elicited long after original infection.

5.8.2 Persistence of memory is ensured either because antigens can persist in native form for long periods of time or because memory B cells tune their signal thresholds and survive in absence of antigen.

5.8.2.1 Thus, memory B cells show different survival requirements and belong in a separate niche that is not dependent on antigen.

5.8.2.2 The number of B cells in the memory pool is controlled by the rates of cell entry, differentiation of activated naïve B cells, and cell exit e.g., terminal differentiation to Ig-secreting plasma cells.

◆◆◆

5.9 TCRs can only interact with peptides associated with MHC.

5.9.1 Thus, T cell activation requires processing of the invader's determinants and their association with the host MHC molecules.

5.9.2 CD4 T cells interact with MHC (class II) molecules that are present in a limited set of the host's hematopoietic cells.

5.9.2.1 The invader determinants associated with MHC class II are captured from the extra cellular environment by the APCs.

5.9.3 CD8 T cells interact with MHC (class I) molecules that are present in all cells of the host.

5.9.3.1 The invader determinants associated with MHC class I derive from intracellular invaders and are synthesized by the infected cell.

◆◆◆

5.10 CD8 T cells participate in IR by eliminating host cells infected by intracellular invaders, e.g., virus.

5.10.1 They interact with peptide of the invader presented by MHC class I molecules of the host.

5.10.2 Upon activation, antigen-specific CD8 T cells proliferate extensively due to the increased availability of resources released by the infected cells and the resulting inflammation.

5.10.3 Among the diverse clones responding to the multiple determinants of the invaders, some are more represented or proliferate faster.

5.10.3.1 Engagement of the antigen specific receptors will confer a competitive advantage to the cells with higher avidity for antigen by increasing their resource usage competence.

5.10.3.2 These processes allow the establishment of dominant T cell clones reacting to "dominant" determinants.

◆◆◆

5.11 Activation induces distinct patterns of genes coding for function, giving rise to different CD8 T cell types.

5.11.1 The "random" gene co-expression patterns in each cell of a population of identical CD8 T cells gives rise to distinct responding cell types.

5.11.1.1 Some promote inflammation, some kill infected cells and others acquire new survival requirements.

5.11.2 CD8 cells producing inflammatory molecules that recruit new cells and enhance IR appear early after activation.

5.11.3 CD8 T cells that kill infected target cells appear at the peak of the response.

5.11.3.1 To kill, CD8 T cells need to co-express different effector molecules: (i) some perforate the membrane of the target cell; (ii) others trigger a cell death program once inside the infected cell.

5.11.3.2 Because the probability of a CD8 T cell to co-express different killer molecules is lower than its probability of expressing a single pro-inflammatory cytokine, cytotoxic T cells appear later during responses.

5.11.3.2.1 Implying that amplifying inflammatory responses allowing entry of diverse new cell specificities precede specific killing of targets.

5.11.3.3 CD8 T cells can only kill target cells that express the same peptide/MHC complex that originally activated the CD8 T cell.[23]

5.11.4 A fraction of activated CD8 T cells acquires new survival requirements.

◆◆◆

[23] Described as MHC-restriction by R. Zinkernagel and P. Doherty.

5.12.1 In an effective IR, after the initial expansion of CD8 T cells, resource and antigen levels drop and the responding clones undergo a phase of contraction.

5.12.2 During the contraction phase cell numbers adjust to the carrying capacity of each clone.

5.12.21 Although dominant and non-dominant clones may share available niches, the variety of resources and the heterogeneity of niches allow co-existence of diverse dominant and non-dominant clonotypes among the persisting cells.

5.12.3 CD8 T cells that persist at the end of the contraction phase are the cells that have tuned their threshold signals and modified their resources requirements.

5.12.31 Upon re-infection with the same invader, these cells will mount a faster and more efficient secondary "memory" IR.

5.12.4 Random activation and plasticity of selected gene expression allow differentiation of activated CD8 T cells to effector and memory subsets.

5.12.4.01 During differentiation, epigenetic changes[24] modify the accessibility of gene regulatory elements increasing the probability of transcription of T cell effector molecules.

5.12.4.1 Successive activation events reinforce gene expression patterns that become stable,

---

[24] Epigenetics refers to changes in gene expression caused by mechanisms other than changes in the underlying DNA sequence.

5.12.4.2 allowing memory CD8 T cells to show more rapid and efficient differentiation to effector functions.

5.12.4.3 Secondary cells recovered after antigen elimination are more efficient killers than cytotoxic T cells present at the peak of the primary response.

5.12.4.4 Thus, while activation of naïve T cells results in a high probability of cells with mutually exclusive effector functions, reactivation of memory cells leads to a higher probability of recovering multifunctional cells, each individual cell expressing different effector functions simultaneously.

5.12.4.5 "Memory" CD8 T cells show an increased division rate and a lower loss rate.

◆◆◆

5.13 In contrast to antibody responses, T cell responses do not undergo affinity maturation.

5.13.1 In contrast to B cells, memory T cells do not mutate their TCRs.

5.13.1.1 BCR somatic mutation may increase B cell avidity for antigen and favor their selection in the presence of limiting amounts of antigen by giving to the mutated cells a survival advantage within the germinal center.

5.13.1.2 In contrast, TCR mutation could affect MHC recognition and, since T cell survival requires interactions with ubiquitous MHC molecules, most likely will lead to cell death.

5.13.2 Memory CD8 T cells follow a different strategy. They undergo a process of functional maturation.

5.13.2.1 In response to recurrent signals, they reinforce gene expression patterns and become ready to express different effector functions simultaneously.

5.13.2.1.1 They are functionally more efficient.

5.13.2.2 They also tune their survival requirements, modify their avidity for antigen and/or MHC, and lower the thresholds of survival and activation.

5.13.2.3 As a consequence, memory CD8 T cells can survive in the absence of the original antigen.

5.13.2.4 The lower signal thresholds allow memory CD8 T cells to exploit weak affinity interactions with different determinants or MHC molecules unrelated to the original activating peptide.

5.13.2.5 Memory CD8 T cells also modify their response to cytokines and acquire the capacity to exploit new resources, such as IL-15.

5.13.3 Memory CD8 T cells inhabit an independent and well-defined niche in the IS, which they share with other populations of activated CD8 T cells.

◆◆◆

5.14.1 While naïve T cells persist mainly as resting cells, memory T cell survival is accompanied by cell division.

5.14.1.1 Indeed, due to their lower thresholds for activation, memory cells are in a pre-activated state, which favors their rapid and efficient engagement in secondary responses.

5.14.1.2 However, and in spite of cell division, the number of memory CD8 T cells is controlled independently of the number of naïve CD8 T cells.

5.14.1.3 During its lifetime, the body generates new populations of memory CD8 T cells in response to diverse invaders.

5.14.2 The IS must accommodate both old and new memory cell populations.

5.14.2.1 Competition based on the receptor specificity and environment heterogeneity may lead to the substitution of the initial memory population by new memory cells.

5.14.2.2 The carrying capacity of each memory T cell clone that is the number of cells that the host environment can support determines the degree of co-existence of the old and new memory cells.

5.14.2.2.1 Attrition of old memory cells is determined both by the rate of proliferation of the new cells and the rate of extinction of established cells.

5.14.2.2.2 Attrition can be induced by factors released during new immune responses to facilitate integration of new memory cells.

5.14.2.3 However, a population of established memory cells may benefit from the founder advantage "first come, first served". In this case it may resist replacement by a new incoming cell.

5.14.2.4 The balance between these mechanisms is determinant for old memory persistence.

5.14.2.4.1 However, as the host is modified by the history of successive invasions, the increased diversity of the resources, their increased levels and varied localization, provide the IS with the plasticity to at least partially accommodate both old and new memory cells.

5.14.2.4.1.01 Homeostasis results from the interactions of a cell assembly with its immediate environment, it varies for each different cell assembly and is therefore adaptable to changes in the environment.

5.14.2.4.1.02 This confers to the IS an adaptive capacity that may allow the accumulation of persistent high numbers of effector/memory cells in particular anatomical sites.

5.14.2.5 Clones bearing TCRs with a broader shape space of recognition have a competitive advantage and have a higher probability of resisting substitution by new cells and of replacing old cells.

5.14.2.5.1 This competitive advantage of the clones with a degenerate specificity allows the IS to keep a broader diversity of recognition in memory T cell compartments.

5.14.2.6 During an IR, the increased production of resources may also lead to by-stander activation of old memory T cells with unrelated specificity and a lower threshold of activation.

5.14.2.6.1 It reactivates "memory" cells from old antigenic experiences—original antigenic sin.

◆◆◆

5.15 Although CD8 T cell IRs can occur in the absence of other lymphocyte populations, CD4 T cells help amplifies CD8 responses and is critical for the full development of memory CD8 T cells.

5.15.1 Helper CD4 T cells express co-receptors and produce factors that favor CD8 T expansion and function.

5.15.1.1 Helper CD4 T cells produce IL-2 that is required for CD8 T cell proliferation and memory differentiation.

5.15.2 CD4 T cell help for CD8 T cells requires close proximity interactions, which may occur at the surface of third party cells the APCs.

5.15.2.1 Activated CD4 and CD8 T cells produce chemokines.

5.15.2.1.1 The CD4/CD8 close vicinity interactions at the surface of a APC also requires the integrity of their chemokine receptors to allow their mutual attraction.

5.15.2.2 While interacting with APC, CD4 T cells induce their maturation and the release of cytokines that facilitate lymphocyte expansion.

5.15.3 At the surface of the APC CD8 T cells interact with peptides of the invader associated with MHC class I, while CD4 T cells interact with peptides from the same antigenic structure associated with MHC class II molecules.

5.15.3.1 CD4/CD8 T cell interactions also involve direct engagement of specific receptor-ligand pairs expressed at the surface of CD4 and CD8 cells.

5.15.3.2 In general these processes contribute to improve the burst size of the responding CD8 T cells and determine their differentiation to the memory/effector stage.

◆◆◆

5.16 Inflammatory cytokines released either directly by the CD4 T cells or by the APCs upon interaction with CD4 T cells further contribute to the amplitude of both primary and secondary CD8 T cell responses.

◆◆◆

5.17 CD4 TCRs interact with antigenic determinants presented by MHC (class II) molecules.

5.17.1 After activation, naïve CD4 T cells proliferate extensively.

5.17.1.1 At the same time they differentiate into distinct effector T-cell subsets, each characterized by their unique cytokine expression and function.

5.17.1.2 Antigen-specific CD4 T cells randomly activate cytokine genes, which implies that within a clone of cells bearing the same antigen receptors each cell may follow a different outcome.

5.17.1.2.1 This is followed by modifications of the accessibility of cytokine gene regulatory elements that involves different and progressive events affecting either the gene or any of its multiple regulatory elements that render the cytokine loci more or less accessible.

5.17.1.3 While in the initial stages of activation, CD4 T cells maintain a flexible cytokine differentiation program; with time and exposure to recurrent activation steps they become polarized to a fixed pattern of cytokine production.

5.17.1.3.1 When cells initiate gene expression each cell shows major cell-to-cell variations in the transcription rates of that gene. In contrast, when transcription becomes permanent transcription rates stabilize and the activated/memory cell becomes polarized.

5.17.1.3.2 Co-production of some cytokines may not be possible, as they represent end-points of mutually exclusive differentiation programs. Each gene has its own peculiar transcription rate.

5.17.2 The pattern of differentiation and type of effector cell subset is determined by the initial context of T cell activation, the type of APCs, by the microenvironment and by the cells "set-up".

5.17.2.1 Cytokines produced by APC and by CD4 cells themselves modify environments and further bias final CD4 T cell differentiation.

5.17.2.2 Rates of cell division of the different cell subsets also drive to the type of response as minor differences in the initial rates of proliferation can determine dominant responses.

5.17.3 There are as many CD4 T cell subsets as the combinations of cytokines they can produce,

5.17.3.01 each polarized CD4 T cell subset is able to exploit distinct environments.

5.17.4 CD4 T cells that mainly produce the IL-4 cytokine favor B cell survival and promote Ig switch and somatic hypermutation in activated B cells.

5.17.4.1 These T cells increase the amplitude of antibody responses and the development and survival of memory B cells.

5.17.4.1.1 Antigen-specific CD4 T cell help to B cells also involves close proximity interactions between the T and B cells.

5.17.4.1.2 T and B cells recognize distinct determinants of the same antigen at the surface of intervening third party APCs.

5.17.4.1.3 Expression of MHC class II by B cells favors direct T-B cell interactions.

5.17.4.1.4 T cell help also implies interactions between specific receptor-ligand pairs expressed at the surface of the T, B and APC cells involved.

5.17.4.1.5 Soluble factors produced by CD4 T cells help responses of B cells recognizing unrelated antigens.

5.17.4.1.6 Such bystander responses are generally outcompeted in the presence of specific B cell responses, able to exploit new unique resources.

5.17.5 CD4 T cells that produce interferon-γ or IL-17 enlist cells that mediate inflammatory responses.

5.17.5.01 In fact harm to the host can be caused directly by the invader or through the host's own reaction.

5.17.5.1 Inflammation encases pathogens within limited bounds, helping to control their dissemination and contributing to their elimination.

5.17.5.2 Inflammation also increases the levels of available resources, thus enhancing the IR.

5.17.5.3 Inflammatory responses are non-specific and thus destructive to all types of tissues, including the host's, and therefore must be kept under control by the production of anti-inflammatory cytokines.

5.17.5.4 CD4 T cells that secrete IL-10, restrain inflammation and bias CD4 T cell differentiation towards an anti-inflammatory pathway.

5.17.6 CD4 T cells that secrete IL-2 facilitate memory CD8 T cell induction and favor proliferation and survival of Treg cells that control naïve CD4 T cell expansion.

5.17.6.1 By regulating CD4 T expansion and helper effects, Treg cells control IR.

5.17.6.2 The balance between Treg cells and the CD4 T cells that produce IL-2 may represent a feedback loop in the contraction phase of the IR.

5.17.6.2.1 Treg cells exert their effects and are inducible mainly through TGFβ.

◆◆◆

5.18 After the initial phase of expansion, responding CD4 T cell clones undergo a phase of contraction and only a fraction of the antigen-specific cells differentiate to a persistent state.

5.18.1 Upon re-infection with the same invader, these CD4 T cells contribute to a faster and more efficient secondary "memory" IR.

5.18.2 Activated CD4 T cells show modified threshold signals and resource requirements and occupy an independent niche.

5.18.3 Activated T cells express selective patterns of chemokine receptors, which allow them to migrate and return to the initial tissue of priming—territorial memory.

◆◆◆

Chapter 6. Vaccines

6. Upon re-infection with the same invader the host can mount a faster and more efficient "memory" response.

6.1 Since memory responses can be elicited a long time after the clearing of the original infection they are the basis of vaccination protocols.

6.1.1 Vaccination aims to prevent entry and spread of invaders among host populations or to protect populations against their noxious products, e.g., toxins.

6.1.1.1 Vaccination should confer long-term protection.

◆◆◆

6.2 Vaccine design requires deep knowledge of both the invader and the host (in particular of its immune system (IS)).

6.2.1 To know the host and the invader is to study all their possible interactions.

6.2.2 Since invaders evolve multiple different strategies to avert host defenses,

6.2.2.1 vaccine design should aim to offset each specific invader strategy.

6.2.2.1.1 Vaccine design also requires knowledge of the biology of the vectors that spread infection by conveying invaders from one host to another.

6.2.2.1.2 Vectors have evolved strategies to circumvent host reactions to their aggression or ingress.

6.2.2.1.3 Vector strategies indirectly facilitate invader entry and spread.

6.2.2.1.4 Indeed, smooth invasion does not provoke immune reactions.

◆◆◆

6.3 Vaccines against a toxin imply immunization of the host with the neutralized toxin together with factors "adjuvants" that help to promote inflammation, and thus increase the amplitude of the immune response (IR) and facilitate the establishment of an immunological "memory".

◆◆◆

6.4 Vaccines against invaders should precisely mimic natural infection.

6.4.01 Vaccines should allow local increased production of resources to enhance IRs.

6.4.1 They should preferentially use low doses of a safe attenuated (or killed) version of the invader given by the natural route of infection.

6.4.1.1 Since invaders have much faster replicating times than the cells of the host, the attenuated version avoids pathogenic effects and gives time for the cells of the IS to divide and mount an efficient primary IR.

6.4.1.2 Vaccine protocols may immunize against whole proteins of the surface of the invader.

6.4.1.3 However, invaders may disguise their coat or enter an organism protected within an infected cell.

6.4.2 Anticipating responses to specific (peptide) determinants of some of the invader's proteins and using them separately as vaccines, ignores the diversity of host populations, the complexity of host-invader interactions and limits the range of protective host responses.

6.4.2.01 Using modified vectors to immunize against a protein from a different invader is to forget that the determinants in the new vector will not behave as in the original invader.

6.4.2.02 Indeed, different responses will be obtained to the same determinant according to the carrier.

6.4.2.03 These responses will be out of context and most will not preformed as originally planned to protect the host.

6.4.2.1 Interaction of the host with a separate fraction of an invader cannot mimic its interactions with the whole.

6.4.2.2 Similarly, using separate cells of the host (APCs) to present parts of the invader and immunize is to ignore that the IS is a set of interactions rather than a cohort of cells and that the "rules" that determine response or non-response are part of a whole entity.

6.4.2.3 The behavior of isolated cell populations does not necessarily replicate their function in natural settings.

6.4.2.3.01 Once isolated APCs will lose repairs and modify their behavior.

6.4.2.3.1 The role of an APC is sternly dependent on their close microenvironment,

6.4.2.3.1.01 and the initial context of lymphocyte activation, which varies among individuals and from tissue to tissue, decides the type of IR.

6.4.2.3.2 The probability that such specific vaccines will protect a population against the original pathogen is low.[25]

6.4.3 A vaccine should primarily elicit the production of neutralizing antibodies.

6.4.3.1 Only soluble antibodies molecules that interact directly with determinants of the invader can prevent entry and spread of pathogens.

6.4.3.1.1 Cells that kill infected cells can not prevent pathogen invasion and by themselves can only protect the host from intercellular spreading of infection.

6.4.3.1.2 Moreover, cytotoxic cells may kill cells that are essential to the host and may be associated with inflammatory reactions that may be harmful to the host.

◆◆◆

6.5 Due to uncontrolled growth and invasive properties, tumor cells may be considered as invaders.

6.5.1 Therapeutic vaccines have been designed to treat cancer.

---

[25] The first priority and duty of a civilized society is to ensure free health care and education to all. Science and knowledge are too valuable to be restrained by "nonsense" economic issues. Vaccine and drug development should be under the strict control of non-profit organizations.

6.5.1.1 In absence of clear common tumor antigens most of these therapeutic vaccines can only function if they are custom designed for each individual case.

6.5.1.2 They may lead to unwanted aggression against the host.

6.5.1.3 The balance between benefit and cost defines the strategy to follow.

6.5.2 Tumor cells expand and invade due to unique growth and cell-to-cell interaction characteristics that permit them to gain competitive advantage.

6.5.2.1 Precise strategies for tumor therapy should interfere with resource usage by the tumor cell either directly, by modifying resource capture or growth receptor signals, or indirectly, by preventing them to establish or reach their proliferation niche.

6.5.2.1.1 These strategies should preferentially target the intrinsic properties of the tumor cells, that confer them a competitive growth advantages.

◆◆◆

6.6 Antibody diversity and their capacity to interact with all possible 3D molecular shapes can also be used to block or stimulate body components and reestablish tissue homeostasis and thus prevent disease.

6.6.1 "Vaccines" can be used to generate antibodies that control unwanted hormonal levels.

◆◆◆

Chapter 7. Pathology

7 The ability of the IS to recognize infinite molecular forms exists pre-imprinted in the host.

7.1 The cells of the IS are able to interact with an infinite of different molecular forms; thus, they are able to interact equally well with molecular shapes from their host and from invaders.

7.1.1 However, while intruders are generally destroyed and eradicated, host components are "ignored" by the IS.

◆◆◆

7.2 It is said that the purpose of the IS is self/non-self discrimination.

7.2.1 This proposition is preposterous; it arises from a failure to understand the logic of the IS.

7.2.2 Applied to lymphocytes, the concept of self/non-self discrimination is nonsensical.

7.2.2.1 Self/non-self discrimination by lymphocyte receptors is not possible as they interact with all possible molecular shapes whatever their origin.

7.2.2.2 Frontiers between host and invaders are blunt, as many intruders establish symbiotic non-invasive relations with the host, share resources and do not provoke an immune reaction.

7.2.2.2.1 Intestinal flora largely exceeds the number of hosts cells and profit from the host environment.

7.2.2.3 Moreover, hosts change through metamorphosis, puberty, pregnancy, aging and when pregnant, mammalian mothers do not reject their fetuses.

◆◆◆

7.3 In a complex system where a large number of different new cells bearing receptor specificities with a vast degree of degeneracy are continuously generated and where some cells can modify receptor specificity by mutation, the presence of cells capable of interacting with the host components is unavoidable.

7.3.1 Moreover, lymphocytes can only be part of a "state" if they suit their immediate environment, i.e., they interact with host's components.

◆◆◆

7.4 The relevant issue is not self/non-self discrimination but the interactions that determine response or non-response. And the balance between harmful vs. helpful responses.

7.4.01 Best defined as a question of quality control[26] of the IS responses.

7.4.1 The response of an individual lymphocyte is a probability function affected by the levels of binding and cross-linking of the antigen-specific receptors and the engagement of additional co-receptors.

---

[26] Philippe Kourilsky.

7.4.1.1 Under limiting conditions, lymphocytes may not respond following antigen receptor interactions; ligand concentrations may not suffice to trigger a response. Conversely, when in excess, these "resources" may be deleterious for cell survival and blunt IRs.

7.4.2 During development and differentiation, lymphocytes engage in niche demarcation and tune their receptor reactivity to ensure survival and avoid competition.

7.4.2.1 Early immature cells are more sensitive to receptor cross-linking and engage faster cell death programs.

7.4.2.1.1 Thus, "immature" lymphocytes that engage in strong interactions while interacting with host components either fail to survive and are lost from the system, i.e., too much of a resource is harmful.

7.4.2.2 Elimination of all immature cells that interact with host components is an impossible task,

7.4.2.2.1 since most host components will not be present during lymphocyte development.

7.4.2.2.2 Or if present, their quantities would not suffice to eliminate of all specific developing cells.

7.4.2.2.3 Moreover, while high strength interactions may be toxic, low strength interactions with host components may result in the survival and expansion of the interacting cells. Remember, only the fitted cells with repairs within the host's components survive.

7.4.2.3 Alternatively, upon strong interactions with antigen, immature cells may differentiate to functions that preclude host tissue destruction.

7.4.2.3.1 They may become refractory to divide upon new stimuli, but secrete cytokines that sway overall responses.

7.4.2.3.2 They may differentiate into cells that control lymphocyte proliferation and inflammation.

7.4.3 The same rules also apply to individual mature lymphocytes that upon certain conditions adjust their signal thresholds to higher levels, but are still susceptible to similar lenience mechanisms.

◆◆◆

7.5.1 Cell localization and niche segregation are also decisive factors in determining responses or non-responses.

7.5.1.1 When a lymphocyte encounters an antigen in the correct niche and in the presence of plenty of resources it will respond and expand proportionally.

7.5.1.2 This is the case for invaders, which are presented to lymphocytes in the appropriate niche and in the presence of an excess of resources provided by a primary inflammatory reaction.

7.5.1.3 In contrast, encounters with host components frequently occur outside lymphoid organs where limiting resources restrict clonal expansion and aggressive responses.

7.5.1.3.1 "Smooth invaders" that do not modify host tissues and do not provoke inflammation can escape IS's mediated elimination.

7.5.1.3.2 In immune-pathology tertiary lymphoid tissues are common at the site of the lesion providing the environment favorable for local immune reactions.

7.5.1.4 Absence of cell-to-cell cooperation, i.e., lack of T cell help, also curtails strong aggressive responses.

7.5.2 Production of anti-inflammatory cytokines by responding cells may also limit host tissue destruction.

◆◆◆

7.5.3 During thymus development, a subset of CD4 T cells differentiates to regulatory functions.

7.5.3.01 Treg cell differentiation is induced by strong TCR and co-receptor interactions and influenced by the factors present in the immediate environment of the developing cell

7.5.3.02 and it is determined by the expression of the Foxp3 transcription factor.

7.5.3.1 Once mature, Treg cells respond to their receptor agonists by inhibiting clonal expansion of other lymphocyte subsets and restricting IRs. They act mainly by preventing CD4 T cell help and inflammation.

7.5.3.2 Increases in the number of Treg cells suppresses IRs and their elimination re-establishes IRs.

7.5.3.3 Antigen stimulation of mature CD4 T cells in particular environmental conditions may also induce their differentiation to regulatory functions. This process plays a decisive role in the final IRs.

7.5.3.4 Thus, the final outcome of the interaction of the cells of the IS with their specific ligands, i.e., response vs. non-response, is determined by the fine balance and cooperation between the different cell populations,

7.5.3.4.1 and in particular on the relative proportions of Thelper cells and Treg cells that get activated.

7.5.4 Failure of these processes induced by either genetic or infectious factors may cause IRs that are harmful to the host,

7.5.4.1 or deleterious to the IS.

◆◆◆

7.6 In the IS, lymphocyte selection and adaptation implies co-evolution involving several different cell lines.

7.6.1 Different subclasses of lymphocytes and antigen presenting cells represent patterns of co-evolved interactions.

7.6.1.2 This embodies a "state" of connectedness between the diverse components of the system.

7.6.1.2.1 Too little or too much connectedness leads to unstable dynamics and provoke unsolicited aggression.

7.6.1.2.2 A lymphocyte clone with only a few isolated "harbors" may escape the system's control.

7.6.1.2.3 A system where most clones share interactions and resources may favor competition exclusion and the establishment of dominant clones,

7.6.1.2.3.1 which increases the probability of "unwanted" responses.

7.6.1.2.4 Thus, in practical terms, too much connectedness implies that all cells use the same resources leading to competition exclusion.

7.6.1.3 Altogether, unbalanced dynamics escape quality control and may result in deviated auto-aggressive IRs.

7.6.1.4 Out of these patterns of interactions emerges the holistic properties of the IS.

7.6.2 The set of rules that control the multiple interactions between all different cells explain the immune system's development and functions.

7.6.3 The global properties of cellular communities by far exceed the simple properties of the individual cells.

7.6.3.1 Thus, besides the mechanisms that at a single cell level may determine a "non-response", the global organization of the system contributes to avoid aggression to the host.

7.6.4 Quality control of the IRs will be determined by the hierarchical organization and the level of connectedness between the multiple individual constituents of the system.

◆◆◆

7.7 Homeostatic mechanisms that control cell numbers and dictate lymphocyte competition also direct immune responses.

7.7.1 Lymphocyte competition controls the growth of the potentially "harmful" cells.

7.7.2 Considering that the primary goal of the cells of the IS is to ensure their own growth and survival, the diversity and co-existence of multiple cell clones limit the size of each competitor clone.

7.7.2.1 In their flight for survival, in response to competition and through cooperation, lymphocytes use different survival signals to occupy different ecological niches during cell differentiation.

7.7.2.2 Niche differentiation allows the co-existence of different cell types, and guarantees both repertoire diversity and efficient immune responses.

7.7.2.3 Thus, in highly diverse system where all different clones must fit in a limited space, the size of each clone will be necessarily limited, their expansion prevented by the presence of competitors.

7.7.2.4 The size of any potentially detrimental clone will be limited and its hostile effects reduced and thus, controlled.

7.7.3 Conversely, reduced diversity will increase the probability for the expansion and fixation of damaging clones that, in absence of competition, will become dominant and induce pathology.

7.7.4 We conclude that lymphocyte diversity may help to prevent the emergence of large clone sizes, clonal dominance and immune-pathology.

7.7.5 The notion of "horror autotoxicus"[27] must be replaced by the concept of "horror monoclonicus".

7.7.5.1 Indeed, both the tadpole, in which the diversity of IS is restricted to about 10^4 lymphocytes, and humans, that contain about 10^{12} lymphocytes, are able to mount efficient IRs when immunized with any type of foreign antigens.

7.7.5.2 Thus, the question is no longer how large should the repertoire be to discriminate all antigens, which can be achieved by a relatively low number of lymphocytes, but rather what is the minimum size of the repertoire that protects from external invasion and also avoids auto-immune pathology.

◆◆◆

7.8 In this context, polyclonal activation may have a therapeutic role in autoimmune disease, while immune suppression may reinforce the already established auto-aggressive clonal drift.

7.8.1 As heterogeneous populations occupy heterogeneous habitats, the probability of extinction of a species is higher for rare populations.

7.8.01 Thus, partial immune suppression, though useful by damping inflammation, may also lead to preferential extinction of rare specificities decreasing diversity and favoring the maintenance of dominant damaging clones.

---

[27] Paul Ehrlich.

7.8.1.1 Immune suppression may help to control the symptoms of the autoimmune diseases, but generally does not act at the primary causes.

7.8.1.2 In fact, immune suppression may reduce lymphocyte diversity, decrease clonal competition and thus, boost pathology.

7.8.1.3 Complex systems tend to reach steady states. After the cessation of immune-suppression the IS will tend to follow a "strange attractor"[28] dynamic and return to disease.

7.8.1.4 Indeed, once established, autoimmune diseases have a propensity to deteriorate host tissues and to strengthen their dominance.

7.8.1.5 In contrast, polyclonal activation by stimulating multiple different cell clones may increase diversity and clonal competition.

7.8.1.5.1 Thus, it may overcome dominance by aggressive clones by allowing the establishment of a new equilibrium.

7.8.1.6 However, too many resources may, in some cases, lead to overall increase in multiple clonal sizes and pathology. In this case, changes of resource intake, e.g., by modifying the intestinal flora, may be helpful.

7.8.1.6.1 The balance between these two possibilities must be evaluated for each individual case.

7.8.2 At a population level there is an inverse correlation between the incidence of autoimmune diseases and the frequency of infectious diseases, which needs to be clarified.

---

[28] Strange attractor is a state to which a dynamical system evolves.

7.8.2.01 The prevalence of autoimmune pathology is higher in developed countries with low population densities and very cold climates that disfavor germ spread.

7.8.2.02 It is lower in highly populated areas of third world countries.

7.8.3 Finally, there is a direct correlation between the incidence of autoimmune diseases and immune-deficient states, either inborn or induced, following chemo- or radiotherapy.

◆◆◆

7.9 The probability of extinction of a population is related to the area of habitat occupied.

7.9.1 Destruction of wide areas of habitat can be more damaging for the most common species with a wider distribution than for rare species that occupy restricted niches and may therefore escape catastrophic events.

7.9.2 Large structural changes in the IS or in the lymphoid organs will predominantly affect dominant clones.

7.9.3 In health threatening conditions, total irradiation followed by autologous bone marrow transplantation might be indicated.

7.9.3.1 Total irradiation will destroy all lymphocyte clones including dominant and abrogate hostile responses.

7.9.3.2 In complex systems very minor differences in the starting context have later major dynamic consequences—butterfly effects.[29]

[29] This refers to the idea that the flapping wing of a butterfly represents a small change in the initial local conditions that might lead to large-scale alterations in the atmosphere.

7.9.3.3 Thus, it is unlikely that the reconstitution of the IS in a new initial reset context, after irradiation, will mimic the conditions that lead to the initial establishment of autoimmune disease.

◆◆◆

7.10 Lymphocytes may simply follow the evolutionary "Red Queen Hypothesis" and continuously adapt to maintain their survival fitness in the organism they belong to.

7.11 In their flight for survival, lymphocytes may be caught in a Prisoner's Dilemma: either they cooperate, i.e., protect the host and survive, or they defect, that is, fail to protect or attack the host and die.

Chapter 8. Conclusion

8.1 "We often think that when we have completed our study of « one » we know all about « two », because « two » is « one and one ». We forget that we have still to make a study about « and »".[30]

8.1.1 The whole is more than the sum of the parts.

◆◆◆

8.2 Details can be understood, but the global view escapes. "You can't see the forest for the trees".

8.2.1 That is to say: details can be listed, but a general conceptual framework cannot be formulated.

8.2.2 Without concepts, information is vague, ambiguous even misleading.

8.2.2.1 Concepts make facts clear with exact boundaries and projects to the future.

◆◆◆

8.3 Selecting pertinent information and suppressing details permits developing a global view of the IS.

8.3.1 To study immune cells out of their coherent state may prove to be meaningless.

---

[30] After A.S. Eddington.

8.3.1.1 It may be possible to describe some individual properties of the cell, but their contribution to the overall dynamics will be lost in translation.

◆◆◆

8.4 In a highly complex system such as the IS "everything that is not forbidden is compulsory".[31]

8.4.1 Speculating within this framework allows you to put forward new ideas to be tested.

◆◆◆

8.5 "There are, however, dangerous hypotheses. Those that are tacit and unconscious, as we do not know them we cannot abandon them".[32]

◆◆◆

8.6 Moreover, an experiment is the investigation of a phenomenon modified by the researcher.

◆◆◆

8.7 A description of the IS tells us nothing about how the IS really functions. It simply tells us something about the way it is possible to describe it.

◆◆◆

---

[31] A.S. Eddington.
[32] From H. Poincaré.

Chapter 9. The End

9. "What we cannot know we must pass over in silence",[33]

for the moment.[34]

◆◆◆

---

[33] After L. Wittengstein.
[34] "I know not what tomorrow will bring". F. Pessoa.

Tractus Immuno-Logicus: A Brief History of the Immune System
by Antonio A. de Freitas
©2009 Landes Bioscience

Suggested Reading

Aherne WA, Camplejohn RS, Wright NA. *An Introduction to Cell Population Kinetics*. Bath: Edward Arnold, 1977.

Axelrod R. *The Evolution of Co-operation*. London: Penguin Books, 1990.

Begon M, Harper JL, Townsend CR. *Ecology. Individuals, Populations and Communities*. Oxford: Blackwell Scientific Publications, 1990.

Bernard C. *Introduction à L'étude de la Médecine Expérimentale*. Paris: Flammmarion, 1865.

Brown JH, West GB. *Scaling in Biology*. Oxford: Oxford University Press, 2000.

Burnet FM. *The Clonal Selection Theory of Acquired Immunity*. Cambridge: Cambridge University Press, 1959.

Buss LW. *The Evolution of Individuality*. Princeton: Princeton University Press, 1987.

Campbell NA, Rice JB. *Biology*. San Francisco: Benjamin Cummings, 2005.

Darwin C. *The Origin of Species*. London: Penguin, 1859.

Dawkins R. *The Selfish Gene*. Oxford: Oxford University Press, 1976.

Dawkins R. *The Blind Watchmaker*. London: Penguin Books, 1988.

Bonabeau E, Dorigo M, Theraulaz G. *Swarm Intelligence: From Natural to Artificial Systems*. New York: Oxford University Press, 1999.

Dugatkin LA. *Cooperation among Animals*. Oxford: Oxford University Press, 1997.

Dugatkin LA, Reeve HK *Game Theory and Animal Behavior*. Oxford: Oxford University Press, 1998.

Dyson F. *The Origins of Life*. Cambridge: Cambridge University Press, 1985.

Ehrlich P. *Proc R Soc*. 1900; 66:424.

Eigen M, Winkler R. *Laws of the Game: How the Principles of Nature Govern Chance*. London: Allen Lane, 1981.

Endler JA. *Natural Selection in the Wild*. Princeton: Princeton University Press, 1986.

Gleick J. *Chaos: Making a New Science*. London: Viking, 1987.

Greene B. *The Elegant Universe*. Reading: Cox & Wyman Limited, 2000.

Greene B. *The Fabric of the Cosmos*. New York: Alfred A. Knopff, 2004.

Hanski I. *Metapopulation Ecology*. Oxford: Oxford University Press, 1999.

Holland JH. *Emergence: From Chaos to Order*. Oxford: Oxford University Press, 1998.

Hood LE, Weissman IL, WoodWB. *Immunology*. Menlo Park: Benjamin/Cummings Publishing Company, Inc., 1978.

Jacob F. *La Logique du Vivant*. Paris: Gallimard, 1970.

Janeway CA, Travers P, Walport M, Shlomchik MJ. *Immunobiology*. New York: Garland Science, 1999.

Jerne NK. *The Harvey Lectures. Vol. 70*. New York: Academic Press, 1976:93-110.

Kropotkin P. *Mutual Aid: A Factor of Evolution*. 1902.

Margulis L. *Symbiosis in Cell Evolution*. San Francisco: W.H. Freeman and Company, 1981.

May RM. *Stability and Complexity in Model Ecosystems*. Princeton: Princeton University Press, 1973.

May RM. *Theoretical Ecology*. Oxford: Blackwell Scientific Publishers, 1976.

Metchnikoff I. *L'immunité des Maladies Infectieuses*. Paris: Masson, 1901.

Monod J. *Le Hasard et la Nécessité*. Paris: Editions du Seuil, 1964.

Nowak MA, May RM. *Virus Dynamics*. Oxford: Oxford University Press, 2000.

O'Neill RV, DeAngelis DL, Waide JB, Allen TFH. *A Hierarchical Concept of Ecosystems*. Princeton: Princeton University Press, 1986.

Randall L. *Warped Passages*. London: Penguin Books, 2005.

Smith DC, Douglas AE. *The Biology of Symbiosis*. London: Edward Arnold, 1987.

Smith JM. *Evolution and the Theory of Games*. Cambridge: Cambridge University Press, 1982.

Smith JM, E. Szathmàry E. *The Origins of Life*. Oxford: Oxford University Press, 1999.

Tilman D. *Resource Competition and Community Structure*. Princeton: Princeton University Press, 1982.

Tilman D, Kareiva P. *Saptial Ecology*. Princeton: Princeton University Press, 1997.

van Valen L. *Evolution Theory 1*. 1973:1.

Vermeij GJ. *Evolution and Escalation*. Princeton: Princeton University Press, 1987.

Wittgenstein L. *Tratactus Logicus-Philosophicus*. London: Routledge & Kegan Paul, 1922.

T - #0595 - 101024 - C0 - 229/152/8 - PB - 9781587063350 - Gloss Lamination